PSpice for Windows

A Circuit Simulation Primer

Roy W. Goody

Mission College, Santa Clara, CA

Prentice Hall

Englewood Cliffs, New Jersey   Columbus, Ohio

Library of Congress Cataloging-in-Publication Data

Goody, Roy W.
      PSpice for Windows : a circuit simulation primer / Roy W. Goody
            p. cm.
      Includes index.
      ISBN 0-02-345022-3
      1. PSpice for Windows. 2. Electric circuits—Computer simulation.
3. Electronic circuits—Computer simulation. I. Title
TK454.G66 1995
621.3815'01'1353—dc20

94-34322
CIP

Editor: Dave Garza
Production Editor: Christine M. Harrington
Production Buyer: Pam Bennett

This book was printed and bound by Semline, Inc., a Quebecor America Book Group
Company. The cover was printed by Phoenix Color Corp.

 © 1995 by Prentice-Hall, Inc.
A Simon & Schuster Company
Englewood Cliffs, New Jersey 07632

Printed in the United States of America

10 9 8 7 6 5 4 3 2 1

ISBN: 0-02-345022-3

Prentice-Hall International (UK) Limited, *London*
Prentice-Hall of Australia Pty. Limited, *Sydney*
Prentice-Hall of Canada, Inc., *Toronto*
Prentice-Hall Hispanoamericana, S. A., *Mexico*
Prentice-Hall of India Private Limited, *New Delhi*
Prentice-Hall of Japan, Inc., *Tokyo*
Simon & Schuster Asia Pte. Ltd., *Singapore*
Editora Prentice-Hall do Brasil, Ltda., *Rio de Janeiro*

Contents

iv

Part IV — The Field-Effect Transistor

Part V — Special Solid-State Studies

Part VI — Special Processes

Part VII — Analog Communications

Appendices

Index

Preface

I never think of the future. It comes soon enough
Albert Einstein

This text represents a window into the future—a window that is very clear and very bright. By the year 2000, nearly all electronic design, development, testing, and troubleshooting will be *simulated* on a computer. All components and processes will exist only as data bases and mathematical operations. Only at the very end of the product development process will the computer's output be turned into a real device.

This text is designed to ease your way into this amazing new world.

MAJOR FEATURES

Simply stated, this is an introductory text on circuit simulation, and here are its major features:

- It is based on the popular PSpice software from MicroSim Corporation.

- It combines both circuit simulation and electronic theory.

- It is designed to supplement or replace a laboratory or theory text in a conventional devices and circuits course.

- It is comprehensive, covering nearly every feature of PSpice that is available.

- Its emphasis is on devices and circuits, with an introduction to DC/AC. (A follow-on, more advanced text covers operational amplifiers, digital, and advanced filter design.)

- It is an introductory text , requiring no previous knowledge of circuit simulation or devices and circuits theory.

- It is aimed at the technology student, but is entirely appropriate for technicians or engineers as well.

PSpice for Windows

THE SOFTWARE

PSpice for Windows is part of *The Design Center CAE* (computer-aided engineering) system from MicroSim Corporation. It provides a fully integrated environment to *capture* analog/digital circuits directly on the monitor, *simulate* the circuit action, *analyze* the results in graphical form, and prepare the design for PC board development. It is incredibly powerful, easy to learn, and simple to use. Quite simply, the *Design Center for Windows* software package is one of the best learning tools to come along in many years.

Fortunately, for those of us in education, MicroSim Corporation has made available an *evaluation* disk at no cost. Copying of this evaluation disk is "welcome and encouraged." *All the activities in this book are based on the evaluation version.* Its only major limitation is that circuits can be no larger than 20 components and must fit on a single schematic page. Also, several advanced features (such as *programmable logic synthesis* and *signal integrity analysis*) are not yet included in the evaluation disk.

HOW TO USE THE TEXT

The first seven chapters (Part I) cover simple DC/AC theory. Because the most basic PSpice techniques are presented in Part I, *all students should perform the seven DC/AC chapters.* Material from the remaining 23 chapters can be more selectively chosen to match the emphasis of the class.

When you complete this text, you may wish to continue your studies with the follow-on text—which covers operational amplifiers, digital, and advanced filter design. This advanced text assumes that you have acquired a working knowledge of the PSpice techniques present by this introductory text. The follow-on text can be ordered from the publisher of this book.

A SUGGESTION

Although circuit simulation is the major design and development tool of the future, we recommend that students also receive "hands-on" experience by prototyping actual circuits and troubleshooting with conventional instruments.

One computer-saving approach is to divide students into two or more groups and rotate between PSpice and "hands-on." It is especially instructive to perform the same activity using both PSpice and hands-on, and to compare the two approaches. *In this regard, most of the experimental activities outlined in this text can be performed using either PSpice or "hands-on."*

Another suggestion is to have the seven DC/AC chapters performed in the DC/AC classes. Then, when students enter the solid state (devices and circuits) course, they can immediately begin the solid-state chapters.

PREREQUISITES

Besides the ability to perform simple mathematical operations, the only prerequisite needed is a cursory knowledge of Windows. If you have no previous experience with Windows, the author has found that a brief half-hour summary will provide sufficient background to begin the activities. A brief tutorial on Windows is presented in Appendix E.

THE EVALUATION VERSION

This text is based on PSpice evaluation version 6.0. Newer versions of the PSpice software are constantly being released and the chances are good that they would work with this manual. However, to be perfectly "safe," we recommend the use of version 6.0. (If your system lacks a math coprocessor, you must revert to version 5.3.) Because all diagrams and written material are on disk, the author and publisher have designed this text for fast update.

CREDITS

I would like to give special thanks to Debbie Horvitz of MicroSim Corporation for her careful review of the manuscript. Her many comments and suggestions and her marvelous sense of humor were most appreciated. Of course MicroSim Corporation deserves special credit for making the evaluation disk available at no cost. Their foresight makes it possible for colleges and universities to teach circuit simulation at the professional level without breaking their ever-shrinking budgets.

Many thanks to Copy Editor Lorretta Palagi for her amazing ability to catch errors and to make many useful suggestions concerning content and format.

Finally, I wish to express my sincere gratitude to Production Editor Christine Harrington and Administrative Editor David Garza of Prentice Hall Publishing. Under their careful guidance, the project steadily moved forward and was released right on time.

Thank you for adopting PSpice for Windows; good luck and good success.

Roy W. Goody
Mission College

PSpice for Windows

Introduction

Dreaming is an act of pure imagination
H.F. Hedge

We all have awakened suddenly from a dream, only to be surprised that the images we experienced so clearly were not real at all—they were *simulated* within the brain. So perfect is the simulation under PSpice, that you may occasionally find the need to "wake up" and be reminded that the circuits and components you are working with also do not actually exist.

Because all components exist only as mathematical abstractions, we can mold them into any configuration, perform any test, and display any results that our mathematics and dreams will allow us.

Without a doubt, the kind of dreaming we will do under PSpice will indeed be an act of pure imagination.

THE CIRCUIT ANALYSIS PROCESS

Under *The Design Center* software package umbrella are three major interactive programs: *Schematics*, *PSpice*, and *Probe*.

To design, modify, or analyze a circuit we call on these programs during a four-step process:

1. Draw the circuit under *Schematics*.
2. Select the mode of analysis under *Schematics.*
3. Simulate the circuit under *PSpice.*
4. Display the results under *Probe.*

THE PSPICE FILES

During the circuit simulation process, the PSpice software creates and accesses a number of files. Because an understanding of these files will enhance your appreciation of PSpice simulation, we describe these major files next.

Circuit Files

The first file created is the *Schematics file* (.sch), generated when a circuit that was drawn on the screen is saved.   When the schematics file is analyzed, three new files are generated: the *circuit file* (.cir), the *netlist file* (.net), and the *alias file* (.als). The circuit file (the "master file") contains the *simulation directives* and references to the *netlist*, *alias*, and *model* files. The netlist file contains a Kirchoff-like set of "equations" that lists *parts* and how they are connected. The alias file lists alternative names for circuit nodes, and the model file lists the characteristics of each component.

Library Files

Each *part* listed in the circuit ("master") file has a *model* definition and a *symbol* definition. The model definition is found in the appropriate *library* file (*.lib*), and is a set of behavioral parameters that determine the part's characteristics. The symbol definition is found in another library (.slb), and specifies the geometric shape that will appear on the *Schematics* screen. Most of the part definitions used in a circuit come from libraries that are shipped with PSpice. However, if desired, the user can create custom parts and libraries.

Output and Data Files

When PSpice is run, each simulation directive in the circuit ("master") file specifies the information to be sent to the *output* and *data* files. The output file (.out) is an ASCII file that holds the "audit trail" for the simulation and contains a wide variety of information, including the original netlist, all output variables, and various tables. The data file (.dat) is sent to *Probe*, which uses the binary information to generate plots and graphs within the probe window.

MOUSE CONVENTIONS

Under Windows, most commands can be entered using either the mouse or the keyboard. In this manual, we concentrate on the use of the mouse, although keyboard action can occasionally speed up a process. We assume the use of a standard mouse with left and right buttons. (If present, the center button is ignored.)

The mouse follows an *object-action* sequence.   First you select an object and then you perform an action.

- A single click left *selects* an item.
- A double click left *performs an action.*

Throughout this text, we will adopt the following convention:

- **CLICKL** or **BOLD PRINT** (*click left once*) to select an item.
- **DCLICKL** (double *click left)* to end a mode or edit a selection.
- **CLICKR** (*click right once*) to abort a mode.
- **DCLICKR** (double *click right)* to repeat an action.
- **CLICKLH** (*click left, hold down, and move mouse*) to drag a
             selected item.   Release left button when placed.
- **DRAG** (*no clicks, move mouse*) to move an item.

GETTING STARTED

Write or call MicroSim Corporation and ask for their *Design Center* evaluation disk set, version 6.0.

MicroSim Corporation
20 Fairbanks
Irvine, CA 92718

(714) 770-3022 - Technical support
(800) 245-3022 - Sales

Because MicroSim updates its software twice a year, it is reasonable to expect that a version later than 6.0 may be available. The chances are good that this later version will support this text—but there is no guarantee. To be perfectly safe, we recommend version 6.0. (Be aware that copying of the *evaluation disks* is "welcomed and encouraged" by MicroSim Corporation.)

INSTALLING THE SOFTWARE

1.   Place disk 1 in any available drive (A or B).

2.   From Windows, enter the *file manager*, and **CLICKL** on drive A or B.

3.   **CLICKL** on *setup.exe*, **File**, **Run**, **OK**.

4.   **OK**, to select default *Install Design Center Evaluation Version*.

5.   **OK**, to select default *C:\MSIMEV60*.

6.   **Yes**, to create Design Center group with icons.

7.   **Yes**, to add Design Center directory to path in AUTOEXEC.BAT.

8.   **OK**, go ahead and modify.

9.   **OK**, let C be the true boot drive.

10. **OK**, to examine AUTOEXEC.BAT.

11. **No**, to run DOS SETUPDEV.EXE.

12. **OK**. (Be sure to reboot.)

13. Reboot the computer and enter Windows.

14. Turn to Chapter 1; you are ready to go!

PART I

DC/AC Circuits

In the seven chapters of Part I, we concentrate on the fundamentals of PSpice. For this reason, the test circuits are relatively simple and are limited to the DC/AC category. In these early chapters you will find that every step in the simulation process is presented in detail and nothing is left to your "imagination."

In Part II, when we begin solid-state studies (the primary emphasis of this text), we assume that the major PSpice techniques of Part I have been mastered, and we can concentrate on more advanced circuit concepts and PSpice features.

CHAPTER 1

Introduction to *Schematics* & *PSpice*
Ohm's Law

OBJECTIVES

- To draw a simple DC circuit using the *Schematics* program.
- To analyze the circuit using the *PSpice* program.
- To display the resulting voltages and currents.
- To demonstrate Ohm's law.

DISCUSSION

In this first chapter, we introduce *Schematics* and *PSpice*, the software components that create and analyze circuits. The emphasis is on *Schematics*, the powerful program that lets us build (capture) circuits by drawing them within a window on the monitor. It is the function of *PSpice* to analyze the circuit created by *Schematics* and generate voltage and current solutions.

To study Ohm's law, we turn to the DC circuit of Figure 1.1. As shown, all circuits consist of s*ymbols*, *attributes,* and *connections*.

- A *symbol* is a graphical representation of a schematic element. Each symbol is associated with a *Part Name* and stored in a symbol library. For example, a resistor uses a zigzag line for its symbol, R for its part name, and is stored in symbol library *analog.slb*.

> When a symbol is placed on the schematic, the part name may or may not appear. (For "generic" devices, such as a resistor, the part name does not appear; for "specific" parts such as a 1N750 zener diode, the part name does appear.)

- *Attributes* are unique labels for parts. Unlike the generic part name, attributes are different for each instance of the same part. Attributes consist of two sections: a *name* and its associated *value*.

    For example, parts R and VSRC (resistor and voltage source) of Figure 1.1 automatically display the value portions of their *reference designator* attributes (R1, R2, and V1). Other attributes, such as *DC=+10V* for part VSRC, are added by the designer. Any attribute may or may not appear on the schematic, at the discretion of the designer.

- *Connections* are made primarily by wire segments and bus strips.

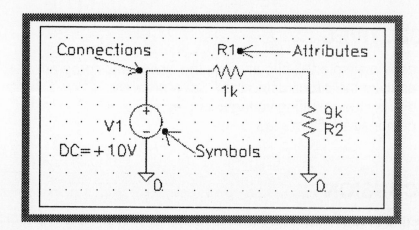

FIGURE 1.1

Simple DC Circuit

BIAS POINT SOLUTION

Most circuits have both steady and changing voltage and current components. Under PSpice, the steady state (DC) component is known as the *bias point* solution and (if present) is always calculated first.

    Because the circuit of Figure 1.1 uses only steady state (DC) voltages, it has only a bias point solution. (During a bias point solution, if capacitors are present, they are opened; if inductors are present, they are shorted.)

SIMULATION PRACTICE

1. As a starting point, we assume that the *Program Manager* window is displayed (Windows 3.1) and the *Design Center Eval* group window is open (see Figure 1.2).

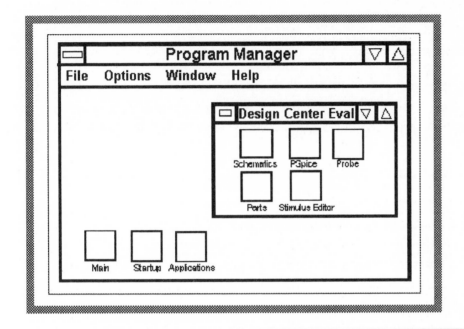

FIGURE 1.2

Program Manager window

Step 1: Draw the circuit

2. **DCLICKL** on the *Schematics* icon within the *Design Center Eval* window and bring up the *Schematics* window (see Figure 1.3).

    This grid of dots forms the work space in which we can draw our circuit. Note the Main Menu bar at the top of the screen (from *File* to *Help)*.

3. To view (pull down) the *File* menu (as shown in Figure 1.3), **File** (remember, this bold print means to click left once on this item). To open a new file, **New**. Note the file name *<new>* in the title bar at the top of the window.

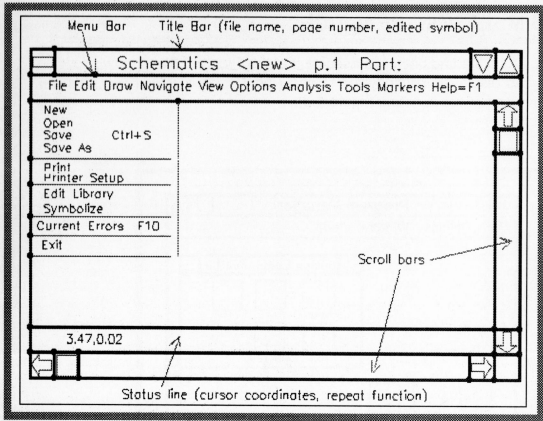

Menu Bar   Title Bar (file name, page number, edited symbol)

Schematics   <new>   p.1   Part:

File Edit Draw Navigate View Options Analysis Tools Markers Help=F1

New
Open
Save   Ctrl+S
Save As

Print
Printer Setup

Edit Library
Symbolize

Current Errors   F10

Exit

Scroll bars

3.47,0.02

Status line (cursor coordinates, repeat function)

FIGURE 1.3

Schematics window
(with File menu
pulled down)

4. To place the first circuit component (the voltage source of Figure 1.4) on the Schematics window, perform the following steps: **Draw** to pull down the *Draw* menu, **Get New Part** to open up the *Add Part* dialog box, **Browse** to open up the *Get Part* dialog box, scroll through *Library* List box (using scroll bar) and **source.slb** when found, scroll through *Part* List box and **VSRC** when found, **OK**, **DRAG** component to desired location, **CLICKL** to place component, followed by **CLICKR** to abort the mode.

5. Note that when placed, the voltage source is automatically *selected* (highlighted red). Review *Schematics Note 1.1* to learn how and why components are selected.

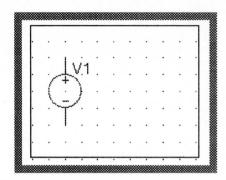

FIGURE 1.4

Voltage source
placed

Schematics Note 1.1
How and why do I <u>select</u> a circuit component?

A component that is selected is red in color.   To select (and highlight red) any circuit
component, CLICKL on the component.   Only a selected component can be operated upon.

6.   The next step is to place both resistors on the schematic as shown
in Figure 1.5. To accomplish this: **Draw**, **Get New Part**, **Browse**,
analog.slb (scroll Library list box), **R** (scroll Part list box), **OK**,
DRAG resistor to R1's location and **CLICKL** (to place R1 and
create R2), **DRAG** second resistor to R2's location and **CLICKL**,
followed by **CLICKR** (to place R2 and abort operation).

Note that all resistors are initially placed horizontally and all
are assigned an initial value attribute of 1kΩ.

7.   To rotate R2: **Edit**, **Rotate**.   (Remember, R2 is highlighted red
and therefore already selected.)

8.   The chances are R2 is now in the wrong location. Reposition R2
(as shown in Figure 1.6), by following the directions of
Schematics Note 1.2.

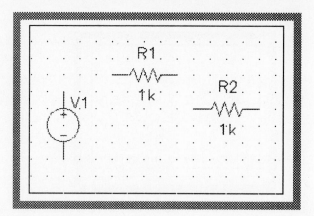

FIGURE 1.5

Initial resistor
placement

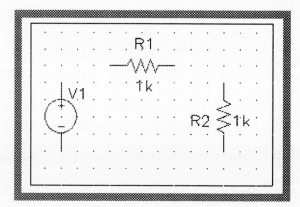

FIGURE 1.6

R2 rotated and
positioned

Schematics Note 1.2
How do I reposition components?

To reposition a component, **CLICKL** to select the component (highlight red), **CLICKLH**
(move cursor arrow over selected component <u>symbol</u> [such as R's zigzag line], hold left
button down), drag to desired location, and release button.

9. To place grounds at the bottom of V1 and R2 (Figure 1.7), perform the following: **Draw**, **Get New Part**, **Browse**, **port.slb** (scroll Library list box), **AGND** (from Part list box), **OK**, **DRAG** ground symbol to bottom of V1, **CLICKL** (to place first ground and create second ground), **DRAG** second ground to bottom of R2, **CLICKL** (to place second ground), **CLICKR** (to abort mode).

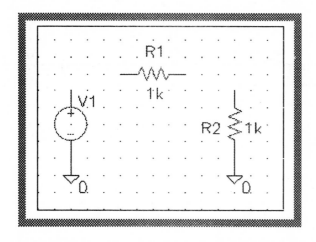

FIGURE 1.7

Placing grounds

10. To complete the wiring, refer to Figure 1.8(a). First, **Draw**, **Wire** to create the "pencil" cursor. **DRAG** the pencil cursor to point 1 and **CLICKL** to anchor the wire to the top of V1, **DRAG** dotted wire straight up to point 2, **CLICKL** to turn wire segment solid and anchor at corner, **DRAG** dotted wire to point 3, **CLICKL** to turn last segment solid and anchor to left of R1, **CLICKR** to abort mode.

   The connection between V1 and R1 should be as shown in Figure 1.8(b). If not, remove wire and repeat (see *Schematics Note 1.3*).

Schematics Note 1.3
How do I delete a component?

To delete any component (including a wire section), **CLICKL** to select (highlight red) the desired component, **Edit**, **Cut** (or press Delete key).

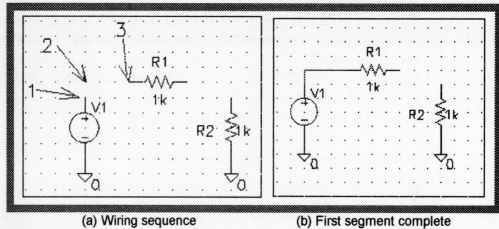

(a) Wiring sequence   (b) First segment complete

FIGURE 1.8

The *Wire* function
(a) Wiring sequence
(b) First segment complete

11. Follow the sequence of step 10 to connect R1 with R2 and complete the circuit (as shown in Figure 1.9).

> Note: To bring back the last command used (shown in the lower right-hand corner after *Cmd:*), **DCLICKR**.

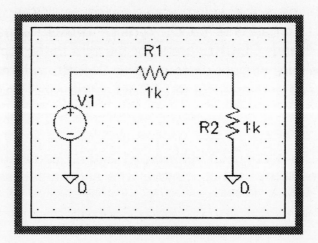

FIGURE 1.9

Circuit wiring
complete

12. The circuit wiring may be complete, but the attributes are not correct.   For example, referring to Figure 1.1, R2's value should be 9k.   Based on the process of *Schematics Note 1.4*, change R2's *value* attribute from 1k to 9k.

Schematics Note 1.4
How do I change an attribute that is presently displayed?

To see how it's done, let's change R2's value from 1k to 9k.

- **DCLICKL** on R2's 1k attribute (not the <u>symbol</u>) to bring up the *Set Attribute Value* dialog box of Figure 1.10.

- Because the 1k value is highlighted in yellow, any entry will automatically delete and overwrite this old value. Therefore, simply enter the new value (9k), **OK**.

  (If you wish to change just a portion of an old value, **CLICKLH** and drag to highlight in yellow any portion of the attribute, and enter the new value. Or, **CLICKL** anywhere inside the entry box to remove the yellow highlight. Then, use the cursor and keyboard to modify the text as desired.)

- On the schematic note that a solid "bounding box" appears about the selected attribute item, and a dotted bounding box appears around the corresponding circuit symbol (to clearly indicate which circuit symbol the new 9k value attribute belongs to). **CLICKL** anywhere on schematic to remove all boxes.

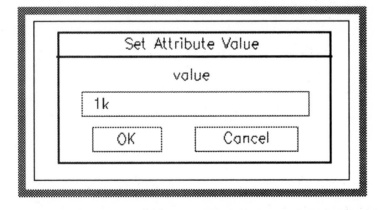

FIGURE 1.10

Set Attribute Value dialog box

13. As shown in Figure 1.1, we must also assign to V1 the voltage level attribute *DC = +10V*. However, this time the process is more complex because this attribute is not presently displayed.

    To add this new attribute, follow the steps of *Schematics Note 1.5*.

Schematics Note 1.5
How do I set and change attributes that are not presently displayed?

As an example, lets assign to V1 the attribute of *DC = +10V*:

- **DCLICKL** on V1's symbol (the circle with a plus and minus inside) to bring up the *Part Name* dialog box of Figure 1.11(a).

- **CLICKL** on *DC=* and fill in the *Value* box with +10V (or simply 10), **Save Attr.** (Do not perform OK yet.)

- The display of any attribute on the schematic is optional. To display the *DC=+10V* attribute on the schematic: **Change Display**, to bring up the *Change Attribute* dialog box of Figure 1.11(b). Within the *Display* box (just above *Layer*), **Value** to enable the display of the +10V value, **Name** to enable the display of the *DC=* name. (Do not perform OK yet.)

  -- Remember, for each attribute, you can display either its name or its value or both.

  -- Although seldom necessary, use the Layer, Orient, Hjust, Vjust, or Size options to customize the attribute's text.

  -- The *Changeable in Schematic* switch gives us an opportunity prevent any attribute from being changed in the schematic. We will leave the default condition (changeable) as is.

  -- The *Keep Relative Orientation* switch is not used.

- **OK** to return to *Part Name* dialog box.

  -- If you wish, **Include Non-changeable Attributes** to enable/disable the listing of the non-changeable attributes (those tagged with an asterisk).

  -- If you wish, **Include System-defined Attributes** to enable/disable the listing of the system-wide attributes (all but *DC=*, *AC=*, and *tran=*).

  When both of the above are enabled, note the list of seven default attribute names, in which *PKGREF* (package reference) and *refdes* (reference designator) have been assigned default values (V1). (A *refdes* value is a unique name for a given part; a *PKGREF* value relates to manufacturing and is not considered within this text.)

- **OK** to return to schematic.

➜ Remember that the display of any attribute name or value (using **Change display**) is optional.

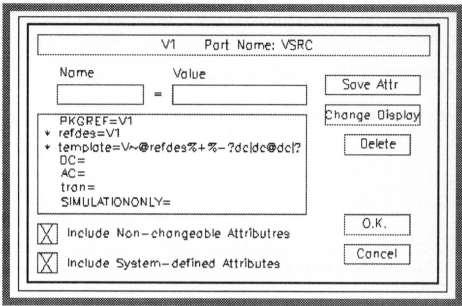

(a) *Part Name* dialog box for source V1.

(b) *Change Attribute* dialog box for source V1

FIGURE 1.11

Attribute-related
dialog boxes
(a) *Part Name* dialog
   box for source V1
(b) *Change Attribute* dialog
   box for source V1

14. The circuit is now complete, but the attributes may not be positioned as desired. To move any of the displayed attributes to a new location (as shown in Figure 1.1), follow the directions of *Schematics Note 1.6*.

Schematics Note 1.6
How do I relocate displayed attributes?

To relocate an attribute to a new location: **CLICKL** on an <u>attribute</u> (not the symbol) to select (surround with solid boundary box), **CLICKLH** and drag to new location, **CLICKL** anywhere on the schematic to remove all boundary boxes.

(Remember from *Schematics Note 1.2*, that the same process is followed to relocate a symbol, except no boundary boxes are generated. When a symbol is relocated, the attributes "tag along.")

---

15. After changes are made to a schematic, the chances are good that the circuit is out of position, incorrectly sized, and that portions of the circuit are fuzzy, splotchy, or missing. After reviewing *Schematics Note 1.7*, *refresh* and *refit* your present circuit, and whenever necessary in the future.

Schematics Note 1.7
How do I "refresh" and "refit" my circuit?

To clarify (refresh) your circuit at any time: **View, Redraw**.
To cause the circuit to fill the screen (fit): **View, Fit**.

Note: When a circuit fills the screen (after **View, Fit**), the circuit automatically moves to allow placement of a new component outside the border. When this happens, **View, Fit** again.

Step 2: Set the bias points

16. The circuit is now complete in all respects. But before we perform the calculations, we must tell the system which voltages and currents to display. Following the guidelines of *Schematics Note 1.8*, place voltage *VIEWPOINTS* and current *IPROBES* as shown in Figure 1.12. (Reminder: *Refit* your circuit as needed.)

Schematics Note 1.8
How do I set up and view DC (bias point) voltages and currents?

For example, to set the <u>voltage</u> viewpoints of Figure 1.12: **Draw**, **Get New Part**, **Browse**, **special.slb** (from Library list box), **VIEWPOINT** (from Part list box), **OK**, **DRAG** to first location (to left of R1), **CLICKL** to place first voltage viewpoint and create second viewpoint, **DRAG** to next location (to right of R1), **CLICKL** to place second viewpoint, **CLICKR** to end mode.

   To set the <u>current</u> probe of Figure 1.12: First use the techniques of *Schematics Note 1.2* to move R2 and its associated ground to the side.   To place the current probe: **Draw**, **Get New Part**, **Browse**, **special.slb**, **IPROBE**, **OK**, **DRAG** to location shown in Figure 1.12, **CLICKL**, **CLICKR**.   Move R2 (and ground) back, as shown in Figure 1.12. **View**, **Fit**, as necessary.

➜ Remember that VIEWPOINTS and IPROBES display bias point (DC) values only.

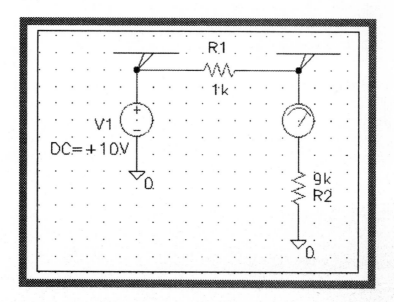

FIGURE 1.12

Setting VIEWPOINTS
and IPROBES

17. Before we begin the circuit analysis process, we must store the circuit file on disk. The first step is to select a disk/directory/file name. Based on file name *ohmslaw*, listed below are several possibilities.

   • **C:\MSIMEV60\ohmslaw.sch** to store in PSpice directory.

   • **C:\MYDIR\ohmslaw.sch** to store in your own directory.

   • **B:\ohmslaw.sch** to store on drive B.

   To store your circuit file to one of the above locations (or other location of your choice): **File**, **Save as** to bring up the *Save As* dialog box. Note the *File Name* box and the *Directories* box. If necessary, **DCLICKL** on the desired drive and directory to place the correct path under *Directories*. Then, enter the desired file name in the *File Name* box, **OK**. (If you omit the ".sch" extension, the system will automatically add it.)

Step 3: Analyze the circuit

18. We are finally ready to perform all calculations and display the results. To accomplish this: **Analysis**, **Simulate**. The *Netlist* window opens during netlist checking, followed by the *PSpice* window during calculations. When finished, "bias point calculated" appears (and a beep sounds).

> When errors occur, either now or in the future, see *Schematics Notes 1.9* and *1.10* to correct the errors and repeat the analysis.

Schematics Note 1.9
How does Schematics help correct errors?

When the errors involve the netlist or electrical rule check (ERC), the *error* window will automatically appear. **OK** to bring up *Error List* dialog box, examine the errors, and perform one of the following:

• **Exit** to return to *Schematics* and correct errors.

• Select error, **GOTO** to have system automatically position cursor arrow at error location on schematic, locate and correct the error.

If no errors are found during netlist\ERC checking, the system enters PSpice. If errors are found during PSpice calculations, we must examine the output file for hints—as explained in *Schematics Note 1.10*.

Step 4: Examine the results

19. Bring back the *schematics* window (**CLICKL** on any window portion that is visible, or use *Alt/Tab* to scroll through the windows). Note the displayed voltage and current values. Are they correct? (Circle "Yes" or "No." If "No," correct errors and repeat.)

    Yes   **No**

20. As an alternative method of obtaining the bias point solution, we may examine the *output file.* To open the output file and analyze its contents, see *Schematics Note 1.10*.

21. Are the voltages and currents listed in the *small signal bias* section of the output file the same as those displayed on the schematic?

    Yes   **No**

22. Whenever you wish to close any or all windows (e.g., you are done and do not wish to perform any *advanced activities* or *exercises*): **File**, **Exit** for each window. (Most windows can remain open until the end of your session.)

Advanced Activities

23. Calculate by hand the total power consumption of the circuit of Figure 1.1 and compare to the value given in the small signal bias solution of the output file. (For this and all future *advanced activities*, report all work and results on separate paper.)

EXERCISES

- Referring to the combination circuit of Figure 1.13, use Kirchoff's laws to calculate (by hand) the current through R2. Verify your prediction using an IPROBE under PSpice. (For this and all future *exercises*, report all work and results on separate paper.)

Schematics Note 1.10
What is contained in the output file?

The output file acts as an "audit trail" of the PSpice simulation and provides many services. For example, whenever an error occurs during simulation that does not involve the netlist or electrical rule check, it is explained within the output file.

The output file is organized sequentially into three major sections: *Schematics Netlist*, *Schematics Aliases*, and the *Small Signal Bias Solution*.

To examine the output file (after running PSpice): **Analysis, Examine Output** from Schematics Window (or **File, Examine Output** from PSpice Window), and scroll through the file.   Based on the circuit of Figure 1.1, the following sections are generated:

<u>Schematics Netlist</u>

```
V_V1    $N_0001 0 DC +10V
R_R1    $N_0001 $N_0002 1k
R_R2    0 $N_0003 9k
v_V2    $N_0002 $N_0003 0
```

Each line of the netlist represents a single component.   For example, the first line uses *reference designator V_V1* to specify a voltage source between node 1 ($N_0001) and node 0 (0) with a dc value of +10V. (v_V2 is the current probe.)   The schematics netlist corresponds to a set of simultaneous equations that PSpice uses to generate the data file.   (The order in which the components are listed corresponds to the order in which they were placed under *Schematics*.)

<u>Schematics Aliases</u>

```
V_V1    V1(+=$N_0001-=0 )
R_R1    R1(1=$N_0001 2=$N_0002 )
R_R2    R2(1=0 2=$N_0003 )
v_V2    V2(+=N_0002 -=$N_0003 )
```

Aliases are useful as alternative methods of specifying nodes.   Each component is assigned special (alias) designators that specify its two ends.   For example, the first line tells us that the top end of component V_V1 is at node $N_0001 and has the alias V1:+.   The bottom end is at node 0 and has the alias V1:-.   The second line tells us that the left end of component R_R1 is at node $N_0001 and has the alias R1:1, and the right end is at node $N_0002 and has the alias R1:2.

<u>Small Signal Bias Solution</u>

NODE	VOLTAGE	VOLTAGE SOURCE CURRENTS		TOTAL POWER DISSIPATION
		NAME	CURRENT	1.00E-02 WATTS
($N_0001)	10.0000			
($N_0002)	9.0000	V_V1	-1.000E-03	
($N_0003)	9.0000	V_V2	1.000E-03	

The small signal bias solution shows the DC voltages at all nodes, all DC circuit currents, and the total DC power dissipation.   To help determine the location of circuit nodes, we may refer to the section on aliases.   The values listed here should, of course, agree with those displayed with VIEWPOINTS and IPROBES.   **File, Exit** to exit the output file.

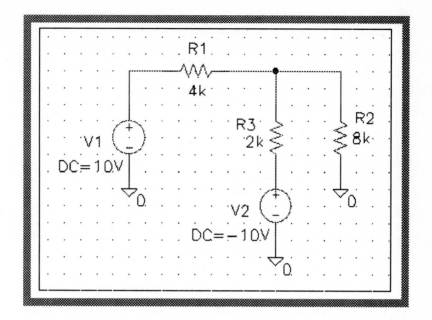

FIGURE 1.13

Applications
circuit

QUESTIONS AND PROBLEMS

1. Write "select item" or "perform action" after each of the following:

    (a) **CLICKL**
    (b) **DCLICKL**
    (c) **CLICKR**

2. How do we know when a particular item on the *Schematics* window is selected?

3. Name two ways of determining a bias point voltage at a node.

4. Part V1 (VSRC) is known as a *voltage source*. What does this mean?   (<u>Hint</u>: What is the output impedance of a voltage source, and how does the voltage depend on the current?)

5. What are the two components of every attribute.   Give several examples.

6. What are the three major sections of the *output file*?

7. What is a *bias point* solution?

8. Is it reasonable that IPROBE be a "perfect" ammeter (one with no internal resistance)?

9. For a resistor, what is the difference between the *part name* (R) and the *reference designator name* (such as R1)?   (<u>Hint</u>: Which one is general and which one is specific?)

CHAPTER 2

Introduction to *Probe*
The DC Sweep Mode

OBJECTIVES

- To prove that a resistor is a *linear* device.
- To determine maximum power transfer.
- To apply the *DC Sweep* mode of operation to DC circuits.
- To use *Probe* to display graphical information.
- To examine and modify a resistor's *model.*

DISCUSSION

Are resistors *linear* devices? That is, is the relationship between voltage and current a straight line—as Ohm's law predicts?

To answer this question, we bring back the test circuit we used in Chapter 1 (reproduced in Figure 2.1). This time, however, we will vary (sweep) the voltage source (V1) across a *range* of values and use *Probe* to display the resulting current on a graph. If the current is a straight line, then resistors R1 and R2 are linear devices.

A second question we wish to investigate with this chapter involves *power*. Again referring to Figure 2.1, what value of R2 will result in maximum power transfer from V1 to R2? To find the answer, we will sweep R2 across a range of values, display the power on a graph, and look for the peak power point.

To sweep V1 and R2 over a range of values we will adopt the *DC Sweep* mode under *Schematics*. We then use the facilities of *Probe* to graph the resulting current and power. (During a DC Sweep, if capacitors are present, they are opened; if inductors are present they are shorted.)

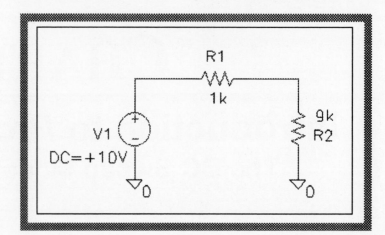

FIGURE 2.1

Simple DC
test circuit

DEVICE MODELS

All the devices used by PSpice rely on mathematical models and model parameters to determine their characteristics. In the case of the resistor (part R), the mathematical model is based on Ohm's law, and the only parameter that can be varied is the *temperature coefficient* (TC).

Looking at the following model equations, we see that ΔR (the change in resistance due to temperature) depends on R (the nominal value of R), ΔT (the change in temperature), and TC (the temperature coefficient):

$$V = I \times (R + \Delta R)$$

$$\Delta R = TC \times R \times \Delta T$$

For example, if the temperature of a 1kΩ resistor changes by 100°C, and TC = .0001, then ΔR = 10Ω. To change the TC parameter value (zero by default), we simply call up the Part Name dialog box and insert a value.

SIMULATION PRACTICE

Problem 1: Are resistors linear devices?

Step 1: Draw the circuit

1. Bring back the circuit of Figure 2.1, used in the previous experiment (**File**, **Open**, **CLICKL** on file name, **OK**).

> Note: If V1 is to be swept, the *DC=+10V* attribute is not necessary. If present (as shown in Figure 2.1) it writes an initial bias point solution to the output file (but only if the *Bias Point Detail* analysis is enabled) and is overridden during the DC sweep process.

2. Following any of the suggestions of *Schematics Note 2.1,* change the position and increase the size of this and any future circuit for easier viewing.

Schematics Note 2.1
How do I change the size and position of my circuit?

To change the size of the circuit, perform any or all of the following:

- **View Fit** to expand the circuit to fill the screen (also see Schematics Note 1.7).

- **View in** (expand) or **out** (contract), **DRAG** crosshairs to desired location, **CLICKL** to increase or decrease size of circuit about the crosshair location.   If you wish, repeat **DCLICKR** to continue the zoom action any number of times.

- To fill the screen with a selected circuit or portion of a circuit: **View, Area, CLICKLH** and drag to create box and perform expansion.

- **View, Entire Page** to show the circuit on a single schematic page.

To change the position of your circuit on the screen, perform any or all of the following:

- **View, Pan - New Center**, **DRAG** crosshairs to desired location, **CLICKL** to reposition circuit about new center point.

- To move the entire circuit or any selected portion of the circuit to any other location, **CLICKLH** at any corner position outside the desired circuit area, drag mouse to create box about circuit.   All circuit components within the box will be selected (highlighted red).   **CLICKLH** anywhere inside the box and drag the box (with circuit elements inside) to new location. **CLICKL** anywhere on screen to remove box.

- We may also change the position of the entire schematic by using the scroll bars.

Step 2: Select the sweep mode

3. To solve our linearity problem, we will perform a DC Sweep of voltage source V1 from 0 to 10 volts.

First, we bring up the *Analysis Setup* menu of Figure 2.2 (**Analysis**, **Setup**) and we enable the DC Sweep mode (**CLICKL** on enable box as shown).   (Do not perform *Close* as yet.)

> If you wish to have a bias point analysis written to the output file (using the *DC=+10V* attribute),   then also enable the *Bias Point Detail* block as shown.

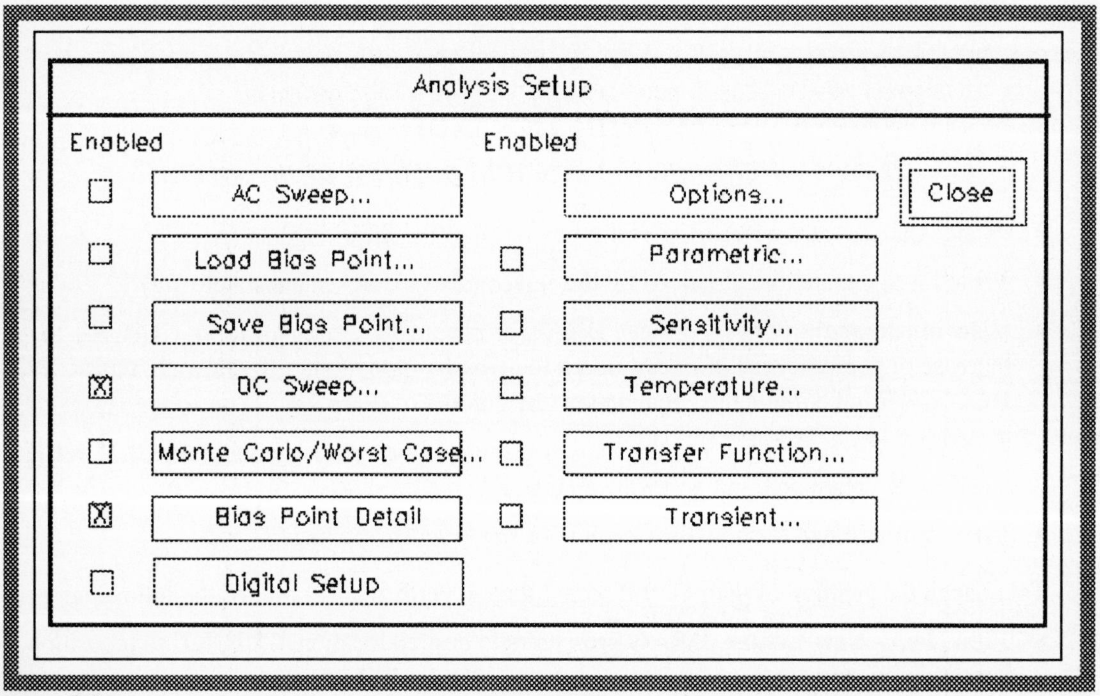

Analysis Setup

Enabled		Enabled		
☐	AC Sweep...		Options...	Close
☐	Load Bias Point...	☐	Parametric...	
☐	Save Bias Point...	☐	Sensitivity...	
☒	DC Sweep...	☐	Temperature...	
☐	Monte Carlo/Worst Case...	☐	Transfer Function...	
☒	Bias Point Detail	☐	Transient...	
☐	Digital Setup			

FIGURE 2.2

Analysis Setup
menu

4. Next, we   **CLICKL** on the *DC Sweep* bar to bring up the *DC Sweep* dialog box of Figure 2.3 and fill in as shown:

    *   *Name:* V1 (The sweep variable.)
    *   *Swept Var. Type*: **Voltage Source**, because we are sweeping V1.
    *   *Sweep Type:* **Linear**, because we wish the voltage magnitude to increase in a linear manner.
    *   *Start Value, End Value, and Increment:* 0, 10, and .1, meaning we wish to sweep V1 from 0V to +10V, in increments of .1V.
    *   *Model Type, Model Name, Param. Name,* and *Values:* not used by this chapter, so leave blank.
    *   *Nested Sweep:* not used by this chapter.

    OK (to exit the *DC Sweep* dialog box and save the parameters), **Close** (to exit the *Analysis Setup* menu with the DC Sweep mode enabled).

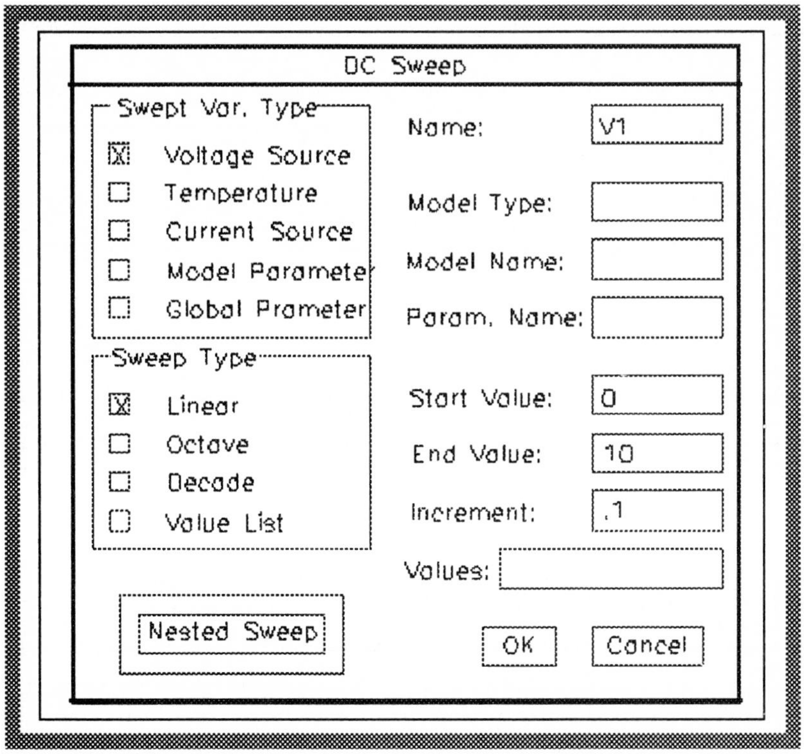

FIGURE 2.3

DC Sweep
dialog box

Step 3: Analyze the circuit

5.   Save the file to disk.   (See Step 17, Chapter 1.)

6.   We begin simulation (**Analysis**, **Simulate**) and the PSpice window automatically appears (Figure 2.4). Note the calculation summary at the bottom, which gives the starting value, ending value, and a running account of the calculations.   (Figure 2.4 assumes that the files are stored in B:\ohmslaw.)

Reminder: If errors are found in circuit wiring, the system will automatically present clues within the *Error List* dialog box.   (**File**, **Current Errors** from the Schematics screen to bring up error box at any time.)   For other types of errors, the system may suggest that we examine the *Output File* for clues (**Analysis**, **Examine Output**).   See *Schematics Notes 1.9* and *1.10*.

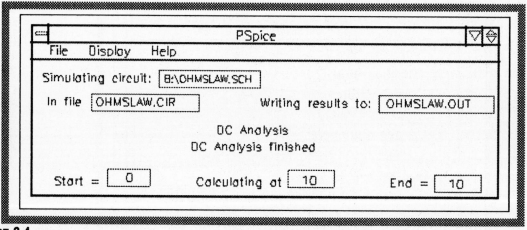

FIGURE 2.4

PSpice window

After completion of the calculations, and assuming no errors were found, the data base generated by PSpice (ohmslaw.DAT) is sent to *Probe* and a graph is displayed (see Figure 2.5).   This is the *initial* (default) graph, in which the X-axis is automatically set by default to the DC sweep variable and range.   As always, the Y-axis is initially blank.

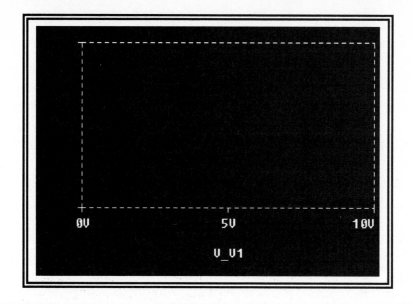

FIGURE 2.5

Initial (default) Probe window showing DC Sweep of V1

Step 4: Select the X- and Y-axis variables

7. To determine the linearity of R1 and R2, convince yourself that the following X- and Y-axis assignments will do the job.

    *   The X-axis should be V1 (V_V1) from 0 to 10V.
    *   The Y-axis should be circuit current.

8. Looking at the *initial* graph of Figure 2.5, the default X-axis variable and range are proper and can remain as is.

9. Bring the *Schematics* window to the forefront (**CLICKL** on any visible portion of the window, or *Alt/Tab* to scroll through the windows.) To set the Y-axis to current and to display the current waveform, we set a *marker* as follows: **Markers**, **Mark Current into Pin**, **DRAG** marker to location shown in Figure 2.6, **CLICKL** (to set marker), **CLICKR** to generate waveform and abort mode. (<u>Caution</u>: Be sure to set current markers at *pin* locations—such as where a resistor and wire meet.)

    Bring the *Probe* graph to the forefront (Figure 2.7) and observe the waveform! (Note that the Y-axis *range* is automatically selected by Probe so the curve will fill the screen.)

PSpice for Windows

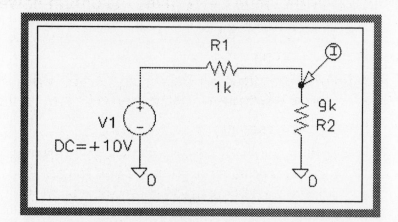

FIGURE 2.6

FIGURE 2.6

Setting a current
marker

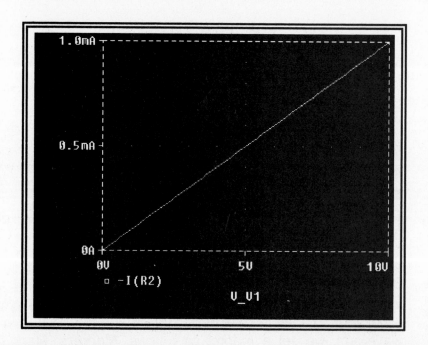

FIGURE 2.7

Graph of current
versus V1

10. Viewing the results (Figure 2.7), can we conclude that a resistor
    is a linear device that obeys Ohm's law?

    Yes   No

Problem 2: What value of R2 gives maximum power transfer?

Step 1: Draw the circuit

11. Remain with the circuit of Figure 2.1. If you wish, erase the current marker (**Markers**, **Clear All** or select marker, **Edit**, **Cut**).

Step 2: Select the sweep mode

12. To generate a graph of power versus load resistance (R2), we must sweep resistor R2 through a range of values. To set up the system for such a *global parameter* sweep, follow the directions of Schematics Note 2.2.

Schematics Note 2.2
How do I sweep a component value?

To sweep a component value (such as R2 of Figure 2.1), we follow a rather involved three-step process:

1. **DCLICKL** on R2's value (presently 9k) to bring up the *Set Attribute Value* dialog box of Figure 2.8. As shown, delete the 9k and enter {RVAL}, **OK**. (The "RVAL" value can be anything, but the curly brackets are necessary.) Note that {RVAL} appears on the schematic (see *Schematics Note 1.6* to change its position).

2. To *define* all parameters used in the circuit (such as RVAL): **Draw, Get New Part, Browse, special.slb, param, OK, DRAG** "box" to any position, **CLICKL** to place, **CLICKR** to abort. **DCLICKL** on "PARAMETERS" to bring up the *Part Name:PARAM* dialog box of Figure 2.9. **CLICKL** on *NAME1=*, enter *RVAL* (no brackets) in *Value* box, **Save Attr.** **CLICKL** on *VALUE1=*, enter 9k in *Value* box, **Save Attr, OK.** (Note: The 9k value for RVAL will be used only for a bias point calculation—and only if the *Bias Point Detail* analysis has been enabled. The value cannot remain blank.)

3. Bring up the DC Sweep dialog box (**Analysis, Setup, DC Sweep**), enter the items as shown in Figure 2.10 (to sweep R2 from 10ohms to 10kohms in units of 10), **OK, Close.** (Caution: Never include zero in your value range.)

After all three steps, the circuit schematic looks like Figure 2.11 and we are ready to proceed to the simulation step.

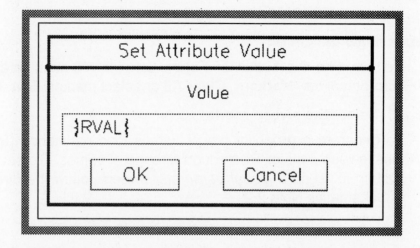

FIGURE 2.8

Set Attribute Value
dialog box

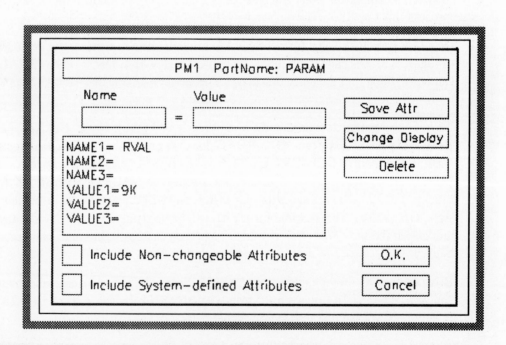

FIGURE 2.9

Part Name param
dialog box

```
                    DC Sweep
┌─Swept Var. Type──────┐   Name:        [RVAL]
│  ☐  Voltage Source   │
│  ☐  Temperature      │   Model Type:  [    ]
│  ☐  Current Source   │
│  ☐  Model Parameter  │   Model Name:  [    ]
│  ▣  Global Prameter  │
└──────────────────────┘   Param. Name: [    ]
┌─Sweep Type───────────┐
│  ▣  Linear           │   Start Value: [10 ]
│  ☐  Octave           │   End Value:   [10K]
│  ☐  Decode           │   Increment:   [10 ]
│  ☐  Value List       │
└──────────────────────┘   Values: [         ]

 ┌──────────────┐          ┌──────┐  ┌────────┐
 │ Nested Sweep │          │  OK  │  │ Cancel │
 └──────────────┘          └──────┘  └────────┘
```

FIGURE 2.10

DC Sweep
dialog box

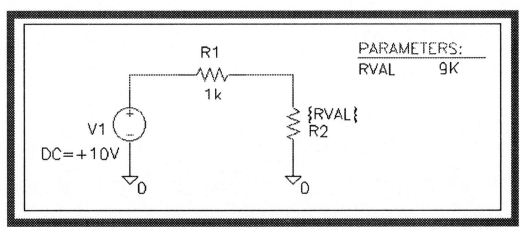

FIGURE 2.11

Circuit set for
R2 scan

Step 3: Analyze the circuit

13. **Analysis**, **Simulate** and generate the default Probe graph of Figure 2.12. (If a warning box appears with "the current data file will be unloaded from Probe," **CLICKL** on OK.)

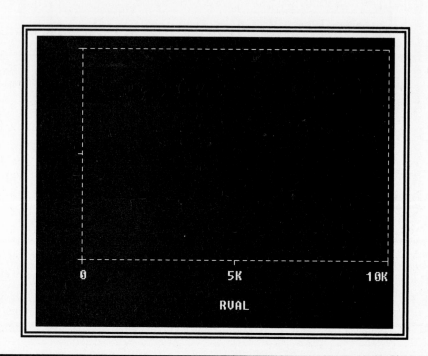

FIGURE 2.12

Default Probe
graph

Step 4: Select the X- and Y-axis variables

14. The default X-axis variable and range are proper. The Y-axis variable is power in R2, which cannot be selected with *markers* (which are limited to voltage and current). To see how power is graphed, follow the instructions of *Probe Note 2.1*.

15. From the resulting power curve of Figure 2.14, *estimate* the value of R2 that results in maximum power transfer.

R2 (maximum power) = _____

Probe Note 2.1
How do I enter custom Y-axis variables?

1. From Probe, bring up the *Add Traces* dialog box of Figure 2.13 (**Trace, Add**). Within the large box at the top is a list of all available *trace variables*. Before moving to the second step, we must understand and properly interpret these trace variables.

   Let's take several examples:

   - <u>V(R2:2)</u>. "V" specifies the variable type (voltage, which is referenced to ground). "R2" specifies a component (such as a resistor), and "2" specifies one end of the component. When any initially horizontal component is first placed, a "1" indicates "left-hand" and a "2" indicates "right-hand." After counterclockwise rotation , node "2" would be at the top.

   - <u>V(V1:+)</u>. This term clearly refers to the positive end of voltage source V1.

     CLICKL on *Alias Names* and observe that additional trace variables are added to the list. These alternative ways of specifying variables are provided for convenience. For example, alias V(R1:1) refers to the same node as V(V1:+) and alias V2(R2) is equivalent to V(R2:2). (**CLICKL** again and the alias names will be deleted.) Also note that we can **CLICKL** on *Voltages* and *Currents* to isolate groups of variables.

2. To create a power variable (or any other custom variable), we must combine appropriate trace variables with mathematical operators. We turn to Appendix B and examine the list of mathematical operators that are available under Probe. Combining the trace variables and mathematical operators, we generate the desired power equation: *Power(R2)* = −*V(R2:2)*I(R2)* [Note: Because I(R2) is negative, the minus sign is needed to generate positive values of power.]

3. To enter the power equation (as shown in Figure 2.13), **DRAG** the "position" cursor to the left-hand end of the *Trace Command* box and **CLICKL** to create the "entry" cursor. Enter "-" from the keyboard, **CLICKL** on V(R2:2), reposition the cursor in the box, enter "*" from the keyboard, **CLICKL** on I(R2), **OK**, and the power curve is drawn (Figure 2.14)!

---

16. If you enabled the *Bias Point Detail* in the *Analysis Setup* window, examine the BIAS POINT SOLUTION of the output file. Do the voltage and current values correctly correspond to R1 = 1kΩ and R2 = 9kΩ?

    Yes   **No**

17. If you wish, follow the directions of *Schematics/Probe Note 2.3* and print a copy of your circuit, graph, or output file.

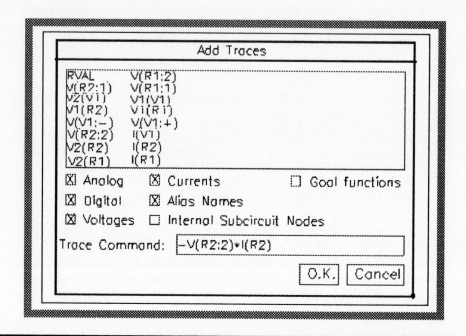

FIGURE 2.13

The *Add Traces*
dialog box

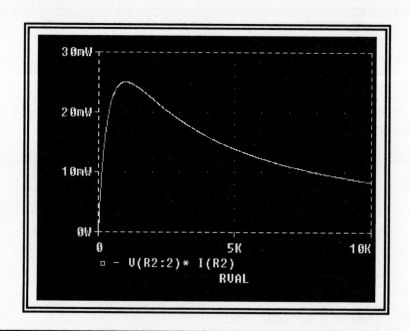

FIGURE 2.14

Power curve for R2

PSpice for Windows

Schematics/Probe Note 2.3
How do I print my circuit, graph, or output file?

Schematics circuits

To print a full schematics page:

- Bring up the *Schematics* window.
- To view a schematics page to see how it will look when printed: **View**, **Entire Page**. (To return to default view: **View**, **Fit**.)
- If necessary **File**, **Printer Select**, and change as desired.
- **File**, **Print**, **Select All** (enable *user-definable zoom factor* and set percentage if circuit is to be printed on multiple pages), **OK**.

To print a selected portion of a schematics page:

- Bring up the schematics page
- Draw a box about the portion you wish to print (**CLICKLH** and drag).
- **File**, **Print**, enable *Only Print Selected Area* box, **OK**. (If desired, enable *user-definable zoom factor* and set percentage if circuit is to be printed on multiple pages.)

Probe graph

- Bring up the *Probe* window.
- If necessary, **File**, **Page Setup**, **Printer Select**.
- **File**, **Print**, **OK**.

Output file

- Bring up the *Schematics* window.
- Open the Output File (**Analysis**, **Examine Output**).
- If necessary, **File**, **Page Setup**, **Print Setup**.
- **File**, **Print**, **OK**.

As an alternative to any of these, zoom (where applicable) to increase the desired size, copy to clipboard, and paste to another document.

MODEL PARAMETERS

18. By modifying the present circuit or drawing a new circuit, recreate the simple test circuit of Figure 2.1.

19. As explained in the discussion, the only model parameter that can be changed for part R is the temperature coefficient (TC). In this section, we will sweep the temperature from –50°C to +50°C and plot the total resistance versus temperature.

    To change TC from the default value of 0 to .0001: **DCLICKL** on R1's symbol to bring up the Part Name dialog box, **CLICKL** on TC=, enter .0001 in the value box, **Save Attr.**, **Change Display**, CLICKL on Value and Name, **OK**, **OK**. Repeat for R2. When done, your schematic should resemble Figure 2.15.

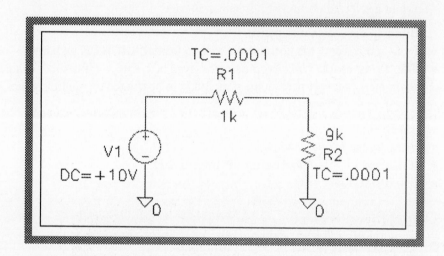

FIGURE 2.15

Temperature
coefficients set

20. To set up the DC sweep: **Analysis**, **Setup**, **DC Sweep**, set up dialog box as shown in Figure 2.16, **OK**, **Close**. (Note: Any item in the *Name* block is not used and can remain or be erased.)

21. Run PSpice (**Analysis**, **Simulate**) to bring up the default Probe graph.   (Is the X-axis temperature from -50 to +50?)

22. Use Ohm's law (R = V/I) to plot the circuit resistance on the Y-axis and generate the graph of Figure 2.17: **Trace**, **Add**, enter – in the Trace Command Box, **CLICKL** on V(V1:+), enter /, **CLICKL** on I(V1), **OK**.

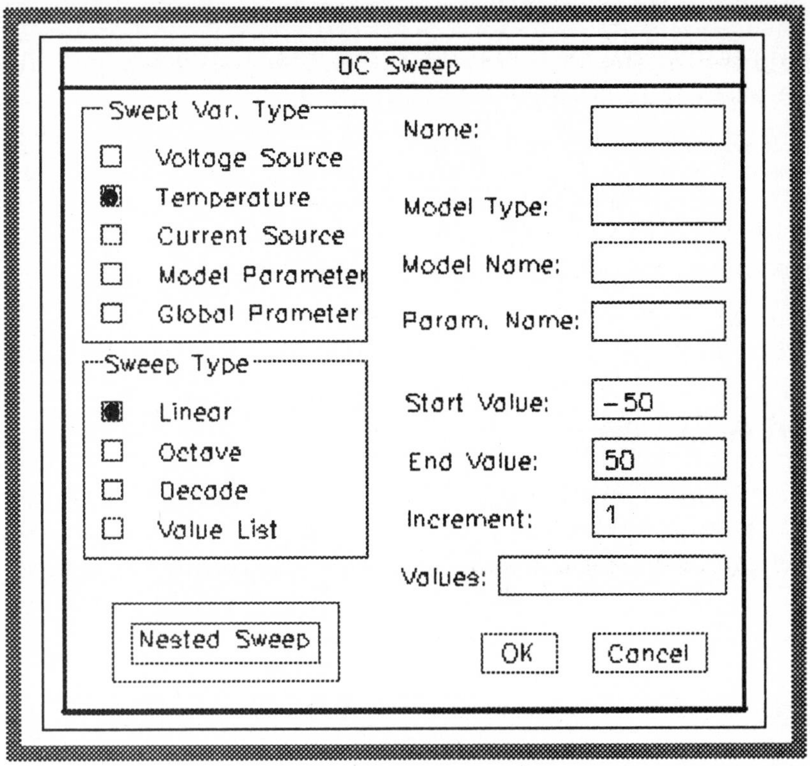

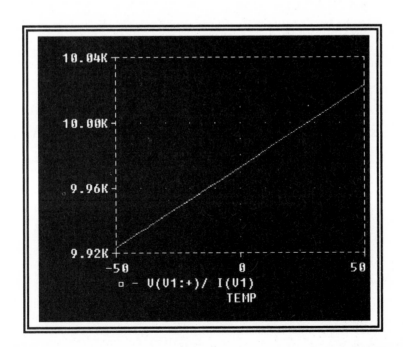

FIGURE 2.16

DC Sweep dialog box for temperature sweep

FIGURE 2.17

Circuit resistance versus temperature

PSpice for Windows

23. Looking at the graph, does the circuit resistance at the nominal temperature of 27.5 appear to be approximately 10kΩ?

   Yes   **No**

24. Does the equation ΔR = TC x R x ΔT appear to hold? (<u>Hint</u>: Is ΔR approximately 100Ω over the 100 degree range?)

   Yes   **No**

Advanced Activities

25. Add the power developed in V1 and R1 to the graph of Figure 2.14. Does P(V1) = P(R1) + P(R2)?   [<u>Note</u>: When using *Probe*, V(R1:1)–V(R1:2) is the same as V(R1:1,R1:2).]

EXERCISES

- Assuming that R2 (the load) can vary from 0 to 10kΩ, use PSpice to determine if R1 and R2 should be a 1/4W or 1/2W resistor. (<u>Hint</u>: What is the *worst case* power dissipated by R1 and R2?)

QUESTIONS AND PROBLEMS

1.   What is the major function of each of the following software components?:

   (a) Schematics

   (b) PSpice

   (c) Probe

2.   Referring to the graph of Figure 2.7:

   (a) How was the X-axis specified?

   (b) How was the Y-axis specified?

3.   Write *voltage marker* or *viewpoint* after each of the following:

   (a) Bias point (DC) voltages on schematic

   (b) Changing (swept) voltages graphed under *Probe*

4.   Based on the results of *Problem 2*, is it true that maximum power is transferred when *impedances are matched*?

   Yes   **No**

5.   *Using current and resistance*, give the custom Y-axis command that would print out the power across R1 of Figure 2.1.

   Power (R1) = _____

CHAPTER 3

AC Circuits
The Transient Mode

OBJECTIVES

- To analyze steady-state sinusoidal AC circuits in the time domain.
- To generate Probe graphs in the transient mode.

DISCUSSION

In this chapter we move from DC to AC circuits. We switch from a constant to a sine wave voltage source, and we adopt the transient (time-domain) mode of display. [The next chapter will investigate the AC sweep (frequency-domain) mode of display.]

We begin with the simple RC circuit of Figure 3.1.

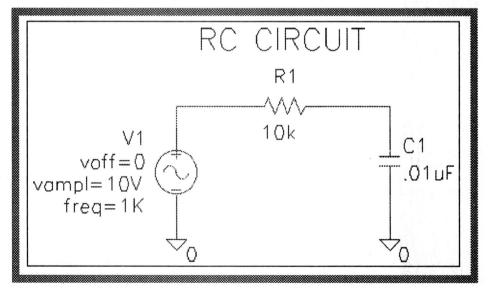

FIGURE 3.1

Simple RC circuit

PSpice for Windows

THEORY

To predict the characteristics of the RC circuit of Figure 3.1, we have the choice of time-domain differential equations or frequency-domain algebraic equations. We choose the following frequency-domain solution because the mathematics is simplified.

Equation		Rectangular	Polar
V1	=	10V	= 10V @ 0°
R1	=	10K	= 10K @ 0°
$-jX_C$	=	$-j(1/(2\pi \times 1k \times .01\mu F)) = -j15.9k$	= 15.9K @ −90°
$I = V1 / (R1 - jX_C)$	=	10 / (10k − j15.9k)	= .53mA @ −58°
$Vout = I \times -jX_C$	=	(10 x −j15.9k) / (10k − j15.9k)	= 8.6V @ −30°

INITIAL TRANSIENT SOLUTION

A transient analysis always begins with an *Initial Transient Solution*, calculated at TIME = 0s. The values used in the calculation are the transient attribute values specified for the source (such as those displayed in Figure 3.1 for sine wave voltage source V1). During calculations, all capacitors are opened and all inductors are shorted. The initial transient solution is written to the output file and is also displayed at the beginning of the Probe plot (at TIME = 0s). The initial transient solution is used as the starting point for a transient sweep.

When performing a transient analysis, we also have the option of generating a bias point solution. This is accomplished by using the *DC=* attribute and enabling the *Bias Point Detail* block in the *Analysis Setup* dialog box. As before, all inductors are shorted and all capacitors are opened. The *Bias Point Solution* is written to the output file, but is <u>not</u> used as a starting point for a transient sweep.

SIMULATION PRACTICE

Step 1: Draw the circuit

1. Draw the circuit of Figure 3.1 and set the attributes as shown.

> V1 has part name VSIN from library *source.slb*. See *Schematics Note 3.1* to set V1's attributes. TD, DF, and PHASE remain at their default values of zero.

Schematics Note 3.1
Which of the 10 attributes of component VSIN do I use?

DCLICKL on V1's <u>symbol</u> to bring up the VSIN *Part Name* dialog box. Note the following list of eight attributes (the non-changeable and system-defined attributes are not listed):

DC=	voff=	freq=	TD=0
AC=	vampl=	PHASE=0	DF=0

- The *DC=* attribute is used only for a *Bias Point Detail* analysis—it's use is optional.

- The *AC=* attribute is used only for a frequency-domain analysis—it is not used when doing a purely transient simulation.

- <u>The remaining six attributes are used for the transient mode</u>. Attributes *voff=* (DC offset voltage), *vampl=* (peak amplitude), and *freq=* (frequency in Hz) must be assigned values. Attribute names *TD* (delay time), *DF* (damping factor), and *PHASE* are assigned 0 by default, but can be assigned other values if desired.

 For each attribute: **CLICKL** on desired name, fill in *value* field, **Save Attr**, **Change Display** (if you wish to display the attribute on the schematic), **OK**.

2. The circuit is now complete. However, to better document the schematic, we wish to add the title "RC CIRCUIT" (as shown in Figure 3.1). This is accomplished as outlined in *Schematics Note 3.2*.

Schematics Note 3.2
How do I add text to my Schematic circuits?

To add text to any circuit: **Draw**, **Text** to open up the Place Text window. Enter text (such as "RC CIRCUIT") in the Text box, change text size as desired (such as 150%), **OK**, **DRAG** text box to desired location, **CLICKL**, **CLICKR**.

Step 2: Select the sweep mode

3. We set the desired *transient* (time) mode as follows: **Analysis**, **Setup**, **CLICKL** on transient enable box, **Transient** to open up the *Transient* dialog box and fill in as shown in Figure 3.2, **OK**, **Close**.

Note: Because the initial time is always zero, <u>we need to specify only the final time.</u> (There is no box for *initial time.*) In our case, a final time of 2ms means that the X-axis of the Probe graph will go from 0 to 2ms and will display two complete cycles of the 1kHz waveform.

As explained in *PSpice Note 3.1*, the default value for *Print Step* (20ns) does not affect the graphical results generated by *Probe* and can remain as is. <u>Do not set the Print Step to zero!</u>

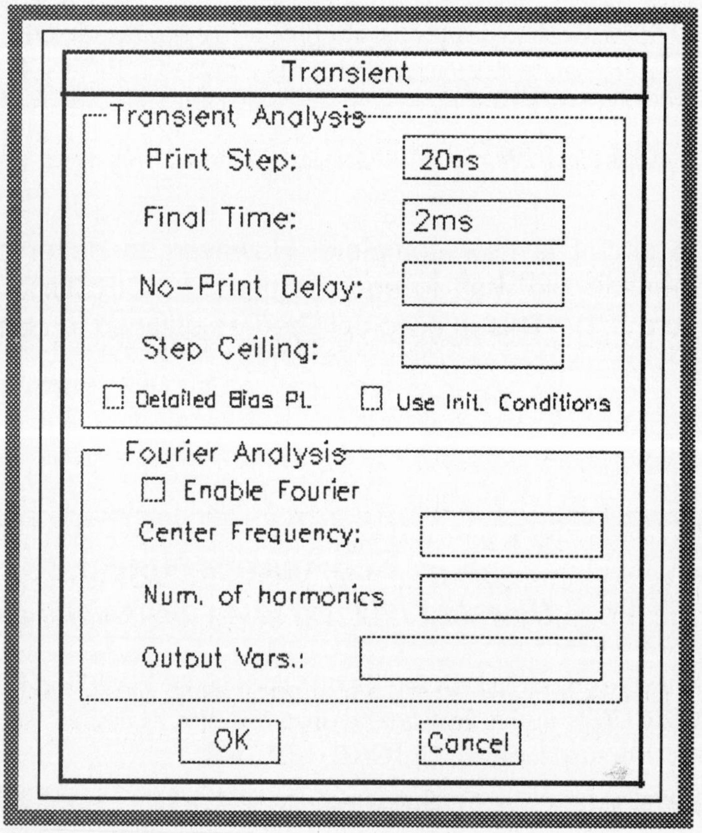

```
                    Transient

    ┌Transient Analysis──────────────────┐
    │                                     │
    │   Print Step:          [ 20ns    ]  │
    │                                     │
    │   Final Time:          [ 2ms     ]  │
    │                                     │
    │   No-Print Delay:      [         ]  │
    │                                     │
    │   Step Ceiling:        [         ]  │
    │                                     │
    │  □ Detailed Bias Pt.  □ Use Init. Conditions │
    └─────────────────────────────────────┘

    ┌ Fourier Analysis───────────────────┐
    │        □ Enable Fourier             │
    │                                     │
    │   Center Frequency:    [         ]  │
    │                                     │
    │   Num. of harmonics    [         ]  │
    │                                     │
    │   Output Vars.:        [         ]  │
    └─────────────────────────────────────┘

        [   OK   ]         [ Cancel ]
```

FIGURE 3.2

Transient mode
dialog box

PSpice Note 3.1
What controls do we have over the transient analysis process?

Referring to the *Transient* dialog box of Figure 3.2, we have the following control over the calculation and display of data points:

- Print Step refers to the print/plot of results to the output file (and not to Probe) when the special PRINT and PLOT *pseudocomponents* are placed on the schematic. Unless these special pseudocomponents are used, *Print Step* can be set to any value that is less than the *Final Time*. (These special pseudocomponents are explained in a future chapter.)

- No-Print Delay is the amount of time (starting from 0) that is not graphed or printed.

- Step Ceiling is used to set the minimum *calculation* interval. All data points will be separated by at least the value of step ceiling. (Step ceiling is useful for reducing the transient calculation time.)

- Detailed Bias Pt. has no effect since the INITIAL: TRANSIENT SOLUTION is always reported by default.

- Use Init. Conditions commands the system to use any individual component initial conditions to generate the INITIAL TRANSIENT SOLUTION. (A future chapter will demonstrate the use of initial conditions.)

Step 3: Analyze the circuit

4. The setup process completed, we save the file (**File**, **Save** or **Save As**, etc.) and evoke *PSpice* (**Analysis**, **Simulate**). (If errors are found, review *Schematics Note 1.9.*)

    When the PSpice window appears (Figure 3.3), note the summary at the bottom, which gives a running account of the *Time step* (time between calculations), the *Time* (present calculation time), and the *End* (finishing time). (To see how PSpice performs transient calculations, read *PSpice Note* 3.2.)

5. After completion of the calculations, and assuming no errors were found, the initial *Probe* window is automatically opened (Figure 3.4).

    A graph appears with the correct X-axis range (0 to 2ms) as specified under the Transient mode dialog box, and with the Y-axis left blank (unspecified).

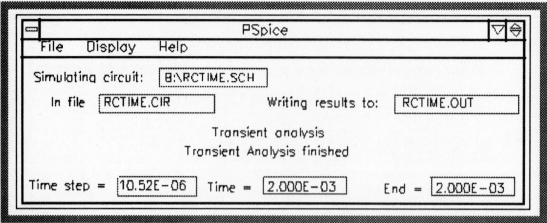

FIGURE 3.3

PSpice Window

PSpice Note 3.2
How does PSpice perform transient calculations?

Each transient calculation is called a *data point*.   During transient analysis calculations, the time step between data points is automatically adjusted. When there is little activity (slow changes), the time step is increased; during busy periods (fast changes), the time step is reduced.

However, regardless of the simplicity of the circuit, at least 50 data points will be computed.   All the data points calculated by *PSpice* are sent to *Probe* for display and printing.

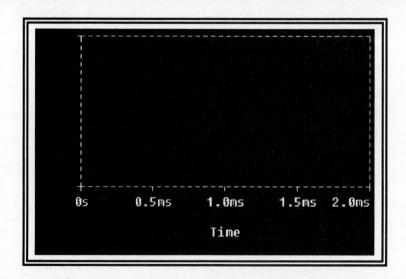

FIGURE 3.4

The initial
Probe window

PSpice for Windows

Step 4: Select the X- and Y-axis variables

6.   The default X-axis variable and range are proper.

To display the input and output voltages as Y-axis variables, we set two voltage "markers" as shown in Figure 3.5 (bring the *Schematics* window to forefront, **Markers**, **Mark Voltage/Level**, **DRAG** marker to first location, **CLICKL**, **DRAG** to second location, **CLICKL**, **CLICKR**)—and the curves of Figure 3.6 appear!

Note: When using markers, the following options are available by using: **Analysis**, **Probe Setup**:

- *At Probe Startup:* Enable <u>Show All Markers</u> to display all marked values automatically upon Probe startup. Enable <u>Show Selected Markers</u> to display only selected (red) markers. (To select multiple markers on schematic, hold shift down and **CLICKL**.)

- *Data Collection:* Enable <u>At Markers Only</u> to collect data only at marked locations (and speed up the analysis).

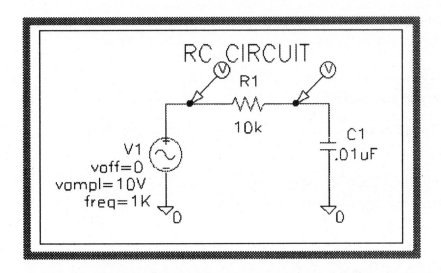

FIGURE 3.5

Setting voltage markers

7.   By viewing the graph (Figure 3.6), *estimate* the following items: (<u>Hint</u>: For determining phase, 1ms = 360°.)

- **Vout(ampl) =** _____

- **Phase difference between Vin and Vout (degrees) =** _____

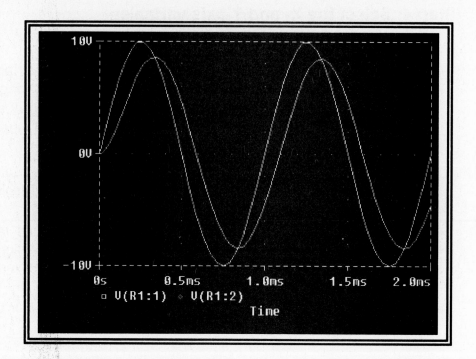

FIGURE 3.6

Probe generated
transient analysis
of RC circuit.

8. In many cases, estimating values graphed under Probe is sufficient (as we did in step 7). However, for those cases in which greater accuracy is required, we use the *cursor*.

    Referring to *Probe Note 3.1* as necessary, use the cursors to again determine the following values as accurately as you can:

    • **Vout(ampl) = _____**
    • **Phase difference between Vin and Vout (degrees) = _____**

9. What is the percent error between the experimental values of step 8 and the theoretical values calculated in the discussion?

    • **Vout(ampl) % error = _____**
    • **Phase difference between Vin and Vout (degrees)
      % error = _____**

Probe Note 3.1
How do I use the cursor to accurately determine waveform values?

To activate the *Probe* cursors: **Tools, Cursor, Display.** Note the appearance of the cursors and the *Cursor* window (see Figure 3.7). If necessary, reposition the Cursor window (using color bar at top of window, **CLICKLH** and drag).

There are two cursors, A1 and A2. (If additional *Probe* windows are opened, both the windows and the cursors take on consecutive letters—A, B, C, ... etc.)

• A1 consists of closely spaced dotted lines and is controlled by the left-hand mouse button.
• A2 consists of loosely spaced dotted lines and is controlled by the right-hand mouse button.

The first step is to associate a cursor with a waveform. This is done by a **CLICKL** (for cursor A1) or **CLICKR** (for cursor A2) on the appropriate color-coded legend symbols in front of the trace variables along the bottom of the graph. For example, on the graph of Figure 3.7, A1 has been associated with V(R1:1) and A2 has been associated with V(R1:2) —as shown below.

$$\boxed{\;\square\; V(R1:1) \quad \diamond\; V(R1:2)\;}$$

To position either cursor, **CLICKL** or **CLICKR** at any graph location and cursor A1 or A2 will move to that horizontal location along the corresponding waveform. The X/Y coordinates of each cursor (as well as the difference) appear in the *Probe cursor* window. For example, referring to Figure 3.7, A1's cursor has an X coordinate of 91.465µs and a Y coordinate of 5.3939 volts.

To fine tune A1 (move A1's position in small steps), we use the arrow keys ($\rightarrow$ and $\leftarrow$). To fine tune A2, we hold down the Shift key and use the same arrow keys. To deactivate the cursor, **Tools, Cursor, Display.**

Plotting current

10. By placing a current marker <u>at a component pin</u> as shown in Figure 3.8 (**Markers, Mark current into pin, DRAG** marker to pin, **CLICKL, CLICKR**), add the circuit current to your graph.

Oops! The resulting graph seems to show a constant current of nearly zero—and we know that is not correct! The problem is that the current *magnitude* is out of range—it's 1000 times less than the voltage magnitude, and PSpice always adjusts its curves according to the largest magnitude curve. (This quite often happens when "mixing" variables, such as voltages and currents.)

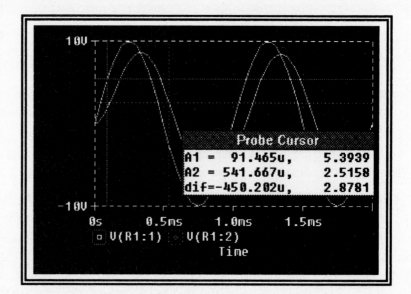

FIGURE 3.7

The Probe cursor
system

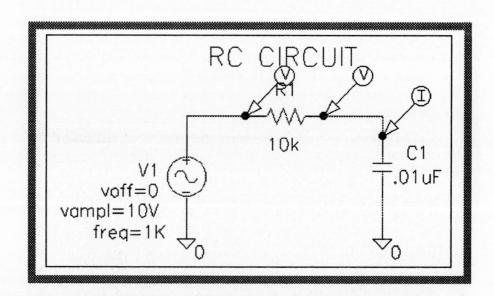

FIGURE 3.8

Placing a
current marker

11. To solve the problem, first delete the present current waveform
     as follows: **CLICKL** on current *reference designator* [-I(C1)] on
     Probe graph to select (highlight red), **Edit**, **Delete**. Or, you can
     select current marker on *Schematics* diagram (**CLICKL** to
     highlight red), **Edit**, **Cut**.

12.   To add back the current waveform correctly, review *Probe Note 3.2* and redisplay the current I(R1) <u>on a new Y-axis</u>. (The result is shown in Figure 3.9.)

Probe Note 3.2
How do I create multiple Y-axes?

If you wish to create a second Y-axis and associate a new waveform with this axis:

1. **Plot**, **Add Y axis** to add new Y-axis to Probe graph.
2. Add desired curve (**Trace**, **Add**, enter trace variable, **OK**, or **DCLICKL** on marker).

   Note the numerical code (1, 2, etc.) at the bottom of the *Probe* window that indicates which waveforms belong to which Y-axis. Also note the "axis marker" (>>) which indicates which Y-axis is active (for example, which axis will receive the next action). To change the active Y-axis (and relocate the axis marker), **CLICKL** anywhere on the appropriate Y-axis scale.

   To delete the selected Y-axis: **Plot**, **Delete Y Axis**.

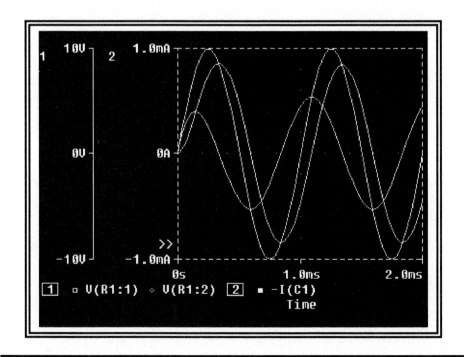

FIGURE 3.9

Adding current to a new Y-axis

PSpice for Windows

13. Using the cursors, determine the following values as accurately as you can. (Use the second positive peak for all measurements, after the system has more closely approached steady state.) How do these values compare with those calculated in the discussion?

- **I(peak) = _____**

- **Phase between Vin and I (degrees) = _____**

14. When a graph contains a large number of curves (as in Figure 3.9), it is convenient to place the legend symbols on the curves themselves. To accomplish this, read *Probe Note 3.3* and reproduce the graph of Figure 3.10.

Probe Note 3.3
How do I place legend symbols on the curves?

- To place legend symbols on the curves: **Tools, Options, Always** (under Use Symbols menu), **OK**.
- To remove the legend symbols: **Tools, Options, Auto** (or **Never**), **OK**.

When in the Auto mode, PSpice automatically places legend symbols when a given graph reaches a certain level of complexity. To permanently change an option, **Save** before **OK**.

To return to default options, **Tools, Options, Reset**.

Initial Transient Solution

15. Examine the output file (**Analysis**, **Examine Output**) and locate the INITIAL TRANSIENT SOLUTION section. Do the initial values match those of Figure 3.9?

      Yes   No

Advanced Activities

16. Perform a power analysis of the circuit of Figure 3.1 by generating the "instantaneous" waveforms of Figure 3.11. By comparing curves, comment on the principle of *conservation of energy* (*power*). [Note: V(R1:2) represents the same node as V(C1:2).]

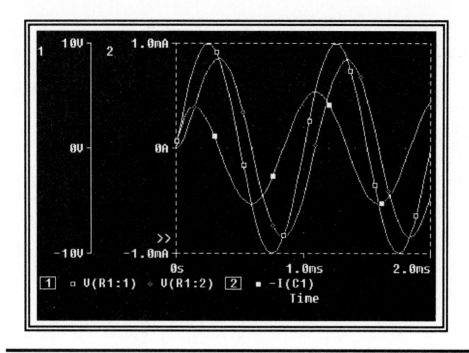

FIGURE 3.10

Using symbols on plots

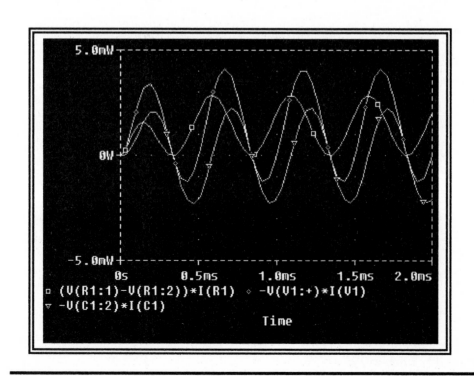

FIGURE 3.11

Instantaneous power curves

17. Use the AVG (average) operator (Appendix B) to add plots of *average* power. (The AVG operator generates a running average waveform.) Again, perform an energy analysis. (Compare the instantaneous power curves of Figure 3.11 with the average curves of this step.)

18. Based on the Circuit of Figure 3.12 (note the added attributes), calculate by hand the INITIAL TRANSIENT and BIAS POINT values and record your predictions on a separate sheet of paper. Also, sketch the expected current and output voltage waveforms. (Be sure to enable the *Bias Point Detail* solution.)

    Generate the same data and waveforms using PSpice and compare with your predictions. (Be sure to examine the output file for the INITIAL TRANSIENT and BIAS POINT solutions.)

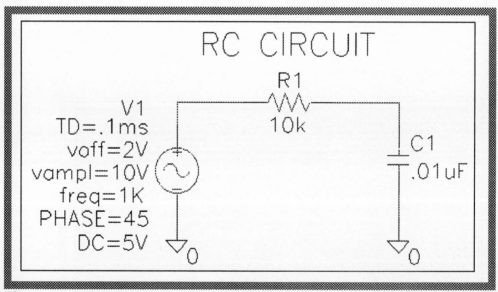

FIGURE 3.12

Attribute test
circuit

19. Our task is completed for this chapter, so we exit all windows (**File**, **Exit**, etc.).

EXERCISES

- Using complex numbers, determine the impedance (Z) of the circuit of Figure 3.13 and convert to polar form. Verify your results using PSpice. (Impedance magnitude is *peak* source voltage divided by *peak* source current, and impedance phase is the angle between peak voltage and current.)

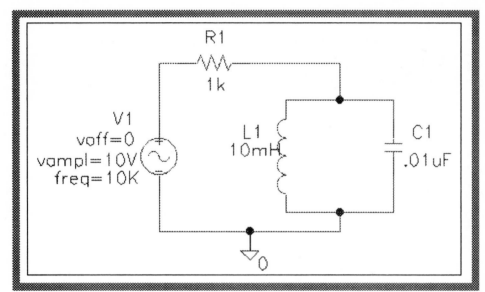

FIGURE 3.13

Applications circuit

QUESTIONS AND PROBLEMS

1. When performing transient calculations, the starting time is always _____.

2. In the complex term –jXc, what does the –j indicate about a capacitor?

3. By drawing directly on Figure 3.1, show the location of the following *reference designators*: V(R1:1), V(R1:2), V(C1:1), and V(C1:2). (Which two are the same?)

PSpice for Windows

4. Referring to Figure 3.11, why is the resistor power always positive and the capacitor power always symmetric about zero? (<u>Hint</u>: What is the difference between *real* and *apparent* power?)

5. If voltage = 10 + j0 and current = 3 + j3, draw several cycles of the current and voltage waveforms on the following graph:

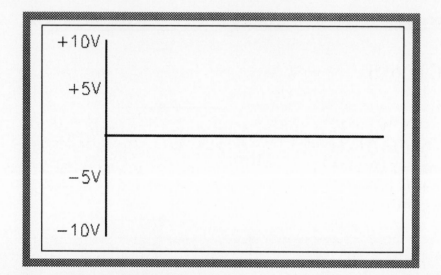

6. What set of sine wave attributes (*Voff, Vampl,* and *Vfreq*) would generate the following input voltage (V1) curve?:

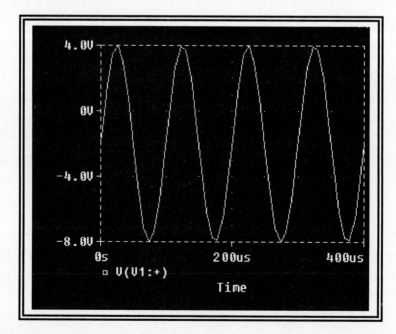

CHAPTER 4

The Tank Circuit
The AC Sweep Mode

OBJECTIVES

- To analyze a parallel RLC tank circuit.
- To generate and analyze Probe graphs in the frequency domain.
- To display complex values in both the polar and rectangular format.

DISCUSSION

One of the most common and useful circuits in all of analog communications is the *tank* circuit of Figure 4.1. It is a natural bandpass *filter*, often used in high-frequency applications to single out a narrow band of frequencies or establish a frequency of oscillation.

When analyzing a circuit, one of the most basic questions we can ask is this: Do we use the *time* domain or the *frequency* domain? The answer often depends on the application at hand. When that application is in the field of analog communications, we are usually interested in a circuit's frequency characteristics—and that requires a frequency-domain analysis.

Frequency-domain analysis means that the X-axis of our Probe display is frequency. Furthermore, it is common to generate *Bode* plots, in which both the X- and Y-axes are logarithmic. To generate a Bode plot, we use the *AC Sweep* mode.

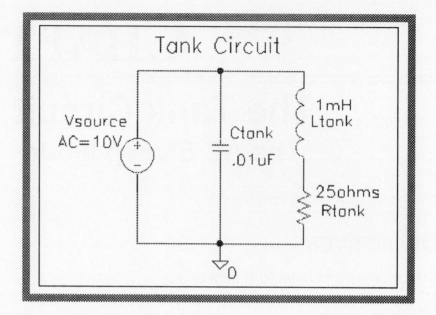

FIGURE 4.1

The tank circuit

THEORY

In the frequency domain, we analyze the tank circuit of Figure 4.1 using Kirchoff's laws and complex numbers:

$$I(source) = I(C\ branch) + I(RL\ branch)$$

$$V(source)/Z = V(source)/-jX_C + V(source)/(R+jX_L)$$

The solution to this equation is quite involved. Fortunately, for small values of R, we can approximate its solution in terms of the tank's *resonant frequency* (f_r) and circuit Q (a measure of the circuit's selectivity). For a high-Q circuit (above 10) the following solutions give satisfactory approximations:

- $f_r = 1/[2\pi(LC)^{1/2}] = 1/[6.28 \times (1mH \times 0.01uF)^{1/2}] = $ **50.33kHz**

- $Q = X_L\ (@\ f_r)/r = (6.28 \times 50.33kH \times 1mH)/25\Omega = $ **12.65**

- BW (Bandwidth) $= f_r/Q = 50.33kHz/12.65 = $ **4kHz**

- $Z\ (@\ f_r) = Q \times X_L = 12.65 \times 316\Omega = 4k\Omega = $ **72dB**

COMPLEX ANALYSIS

All voltage and current variables processed by PSpice are in complex form. By making use of the following *Probe* operators, we can display any circuit value in either polar or rectangular format. (Note that when no operator is used, *Probe* displays the polar format magnitude by default.)

Polar Operators	Description
None (default)	Magnitude of x
P(x)	Phase of x
Rectangular Operators	**Description**
R(x)	Real part of x
I(x)	Imaginary part of x

SIMULATION PRACTICE

Draw the circuit

1. Draw the circuit of Figure 4.1 and set the attributes as shown. (To speed up the drawing process, see *Schematics Note 4.1*.)

> Note: In addition to VSRC, VSIN can also be used for an AC sweep by specifying the *AC=* attribute; however, the transient attributes must also be set.

Schematics Note 4.1
How do I use the keyboard to shorten operations?

For most Schematics operations, we have the option of using the keyboard (instead of the mouse) to shorten the sequence. For example, in Figure 4.2 the *Draw* menu has been pulled down. On the left are the mouse operations, and on the right are the keyboard alternatives.

To take a specific case, *Ctrl+G* performs the same operation as *Get New Part*. Therefore, instead of the consecutive mouse operations **Draw**, **Get New Part** to bring up the *Add Part* dialog box, perform the single keyboard operation: *Ctrl+G*.

A particularly useful keyboard function is *Space* (press space bar), which repeats the last command (which is displayed in the *Cmd:* box at the lower right of the screen.)

Although this text generally uses only the mouse-based commands, feel free to use these keyboard alternatives whenever you wish.

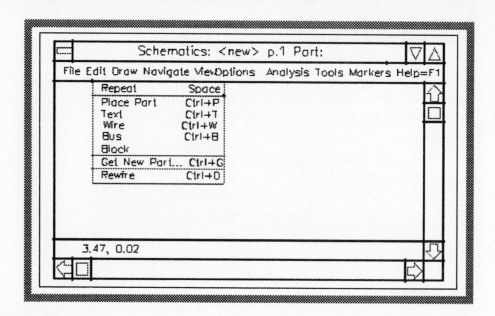

FIGURE 4.2

Using the keyboard
alternatives

Select the sweep mode

2. The sweep variable is *frequency*, to be generated by the *AC Sweep* mode. With a predicted resonant frequency of 50kHz, we decide to sweep the frequency from 10kHz to 100kHz logarithmically. (Be aware that at least one decade of frequency must be specified for logarithmic sweeps.)

    We set up the AC Sweep mode as follows: **Analysis**, **Setup**, **CLICKL** on *Enabled* box for AC Sweep, **AC Sweep** to bring up the *AC Sweep and Noise Analysis* dialog box of Figure 4.3, and enter data as shown.

    *   *AC Sweep Type:* **CLICKL** on *Decade*   (to make the X-axis plot logarithmic).

    *   *Sweep Parameters:* Enter 10 in the *Pts/Decade* box (to command PSpice to calculate 10 points each decade). Enter 10kHz in the *Start Freq:* box, and 100kHz in the *End Freq:* box. **OK**, **Close**.

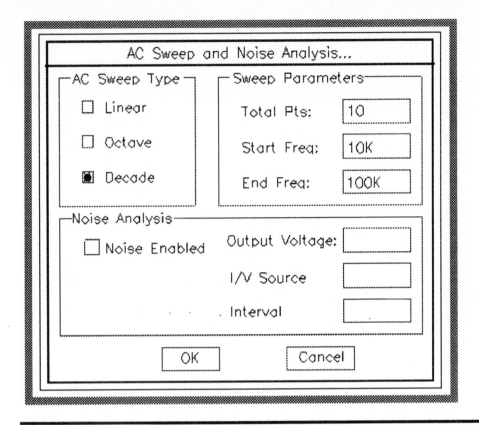

FIGURE 4.3

AC Sweep and
Noise Analysis
dialog box

Analyze the circuit

3. Invoke PSpice (**Analysis**, **Simulate**) and note the calculation
    summary in the PSpice window. When the default Probe graph
    comes up, note the X-axis logarithmic plot of frequency. (Does it
    match your instructions?)

Select the X- and Y-axes variables

4. The default X-axis variable and range are okay. For the Y-axis
    variable we initially choose source current *(I(Vsource))* by setting
    a current marker (Figure 4.4). Is the resulting current plot (Figure
    4.5) reasonable?

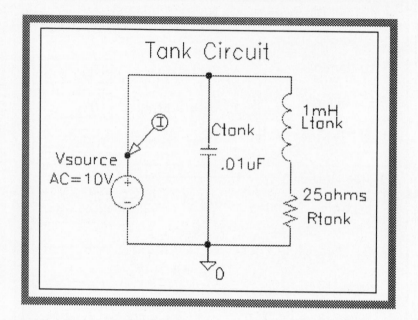

FIGURE 4.4

Showing placement
of current marker

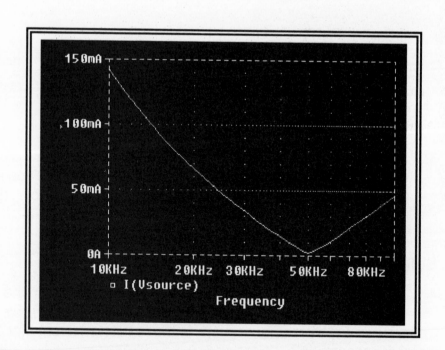

FIGURE 4.5

Plot of current, showing
resonant minimum

PSpice for Windows

5. To switch from linear to logarithmic, we display the current in decibels by taking advantage of the special *advanced markers* available under *Probe.*

    Erase the linear current waveform by deleting the marker from the Schematics window (select marker, **Edit**, **Cut)**. Following the guidelines of *Probe Note 4.1,* place the *idb* marker at the same location and generate the decibel plot of Figure 4.6.

Probe Note 4.1
What special markers are available to plot advanced waveforms?

To use the advanced markers, bring up the Schematics window, **Markers**, **Mark Advanced**, select the desired marker from the following, **OK**, place the marker at the desired location, **CLICKL**, **CLICKR**:

vdb	vphase	vgroupdelay	vreal	vimaginary
idb	iphase	igroudelay	ireal	iimaginary
POLARIS				
IMARKER				
NODEMARKER				
VDIFFMARKER				

- The db, phase, real, and imaginary markers generate the indicated plots.
- IMARKER and VDIFFMARKER are the same as **Mark Current Into Pin** and **Mark Voltage Differential**.
- The groupdelay marker performs the negative derivative of the phase with respect to the frequency (-dPHASE/dFREQUENCY).
- POLARIS is used with *signal integrity analysis* (not covered in this text).
- NODEMARKER is the same as **Mark Voltage/Level**.

---

6. Although the current waveform of Figure 4.6 does show the properties of a tank circuit, we much prefer the traditional plot of *impedance* versus frequency (a *Bode* plot).

    Again, delete the new current waveform (see step 5) , and use the DB (decibel) operator to generate the impedance waveform of Figure 4.7 [Bring up the Probe window, **Trace**, **Add**, enter DB(V(Vsource:+)/I(Vsource)) in the *command box*, **OK**].

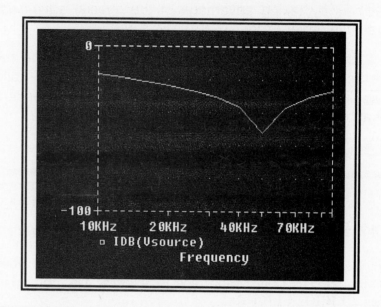

FIGURE 4.6

Plot of current
in decibels

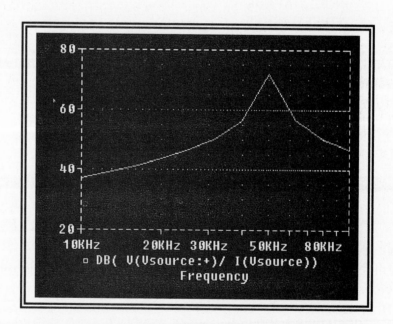

FIGURE 4.7

Bode plot
(dB Y-axis)

7. Based on the waveform of Figure 4.7, use the cursor to determine the following as accurately as you can: [*Bandwidth* is the width of the impedance curve between the 3dB down (70.7%) points.]

$$f_r \quad = \underline{\hspace{3cm}}$$
$$Z(dB @ f_r) \quad = \underline{\hspace{3cm}}$$
$$BW \quad = \underline{\hspace{3cm}}$$
$$Q = f_r/BW \quad = \underline{\hspace{3cm}}$$

8. Do the PSpice values of step 6 approximately match the predicted values calculated in the discussion?

Yes   No

Showing the data points

9. Reviewing Figure 4.7, we look at the very "triangular" shape of the impedance curve around resonance and we wonder if enough *data points* have been calculated to give good resolution.

Fortunately, this is easily answered. Just perform **Tools**, **Options**, **Mark Data Points**, **OK** from your Probe graph, and the data points appear (see Figure 4.8).

10. The result (Figure 4.8) shows that our suspicions were correct! Only three data points define the important resonant portion of the graph.

To provide greater resolution (at the expense of greater calculation time), return to *Schematics* (remove any markers) and bring up the *AC Sweep and Noise Analysis* dialog box (**Analysis**, **Setup**, **AC Sweep**), increase the *Total Pts:* (per decade) from 10 to 50, rerun PSpice, and display the new dB curve. (If necessary, see step 6.)

The result (Figure 4.9) shows a curve of considerably greater resolution. Approximately now many data points now mark the resonant portion of the graph?

Data points @ resonance = \underline{\hspace{3cm}}

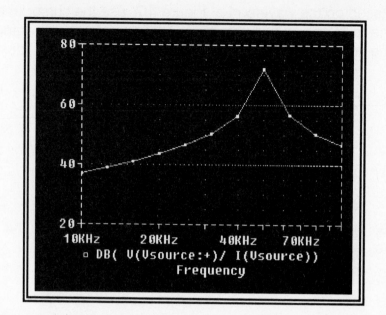

FIGURE 4.8

Graph showing
data points

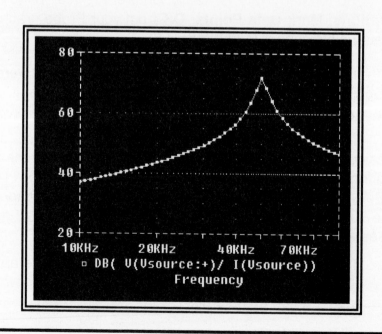

FIGURE 4.9

Increasing the number
of data points for
greater resolution

PSpice for Windows

11. Our resolution study done, we remove the data points to make the waveforms easier to read (**Tools**, **Options**, **CLICKL** on *Mark Data Points* to disable, **OK**).

Complex number analysis

12. To perform a complex number analysis, we plot impedance in both rectangular and polar formats. However, displaying both formats on a single graph would be confusing.

    Based on the techniques of *Probe Note 4.2*, open a "tile" of two Probe windows—as shown by Figure 4.10.

Probe Note 4.2
How do I open up multiple windows under Probe?

To open up a new window under Probe: **Window**, **New** and a new window appears. To arrange the windows in "tile" format (side-by-side): select (**CLICKL**) the window that is to appear at the left, **Window**, **Arrange**, **Tile** (or **Tile Vertical**), **OK**. (**Tile Horizontal** arranges the windows in horizontal layers, and **Cascade** arranges the windows like file folders.)

  As explained in Appendix E, increase the size of the main Probe window as desired. To reconfigure windows at any time, follow the preceding steps: **Window**, **Arrange**, etc.

  To select a given window for activity, **CLICKL** on that window (or **Window**, **CLICKL** on 1, 2, etc.)   To remove any window: **CLICKL** to select, **Window**, **Close**.

---

13. The impedance amplitude shown on the left-hand graph of Figure 4.10 is the polar-form magnitude by default.

    To add the polar-form phase: **CLICKL** anywhere on the left-hand graph, **Plot**, **Add Y-axis**, **Trace**, **Add,** enter

    $$P(-V(Vsource:+)/I(Vsource))$$

    OK. Your resulting graph should resemble that of Figure 4.11, where *d* equals *degrees*. (The minus sign is used to invert the current so the impedance operator phase will correctly lie between $\pm90°$.)

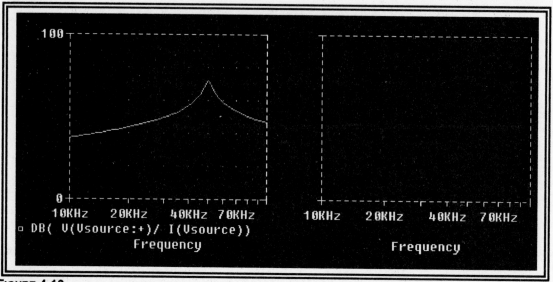

FIGURE 4.10

Initial two-window tile

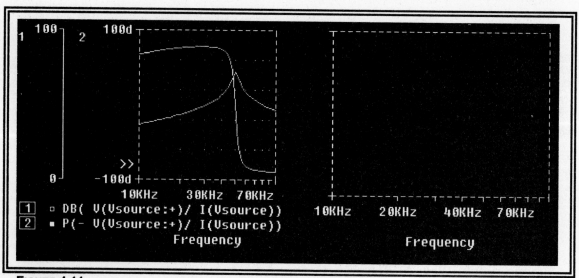

FIGURE 4.11

Adding polar-form phase

PSpice for Windows

14. Next, we place the rectangular-format real (R) and imaginary (IMG) components of the circuit impedance on the right-hand graph.

Select the right-hand graph (**CLICKL** on graph), **Trace**, **Add**, enter the following

R(-V(Vsource:+)/I(Vsource))   IMG(-V(Vsource:+)/I(Vsource))

OK. (Note that we can enter more than one waveform at a time.) Your resulting graph should resemble that of Figure 4.12.

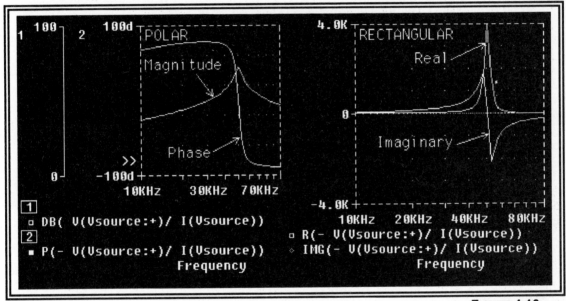

FIGURE 4.12

Final display of Impedance in polar and rectangular form

15. To document your waveforms, as shown in Figure 4.12, see *Probe Note 4.3*.

> To refresh your Probe plots at any time: select window (if necessary), **View**, **Redraw**.

Probe Note 4.3
How do I document my Probe graphs?

If more than one Probe window is open, first **CLICKL** to select the desired window. From the main menu bar on the Probe graph: **Tools, Label** to pull down the label menu. Each of the items listed can be used to document your graph:

- **Text** to bring up the *Text Label* dialog box, enter the desired statement, **OK**, **DRAG** to desired location, **CLICKL** to place.
- **Line** to create "pencil," **CLICKL** to anchor an endpoint, **DRAG** pencil to draw line, **CLICKL** to anchor second endpoint.
- **PolyLine** to create "pencil," **CLICKL** to anchor first endpoint, **DRAG** pencil to draw line, **CLICKL** to anchor end of first line and create beginning of second line, **DRAG** pencil to draw second line, **CLICKL** to anchor second line and create beginning of third line, etc. **CLICKR** when done.
- **Arrow** to create "pencil," **CLICKL** to anchor end of arrow, **DRAG** arrow tip to desired location, **CLICKL** to place.
- **Box** to create "pencil," **CLICKL** to anchor corner of box, **DRAG** to create box, **CLICK** to place box.
- **Circle** to create "pencil," **CLICKL** to anchor center of circle, **DRAG** to create size of circle, **CLICKL** to place circle.
- **Ellipse** to open up "Ellipse Label" dialog box, enter inclination angle (0 or 90 for up/down inclination; 45 for 45-degree tilt to ellipse), **CLICKL** to anchor center of ellipse, **DRAG** to create ellipse size and shape, **CLICKL** to place ellipse.
- **Mark** to display coordinates of last cursor positioned. (Cursor must first be enabled.)

     Edit, Delete to remove any selected (red) labels.

16. Next, we will prove that the polar and rectangular formats of Figure 4.12 are equivalent. The first step is to choose any convenient frequency (such as 45kHz), and use the cursors to fill in the following table.

> Notes: A separate set of cursors must be opened for each window. **CLICKL** on any window to select. Be aware that you may not be able to select all four points at precisely the same frequency.

From Polar Plot	From Rectangular Plot
Magnitude = _____	Real = _____
Phase = _____	Imaginary = _____

17. The next step is to enter the rectangular data of step 16 into the conversion equations below and generate a polar result. Does the calculated polar result *approximately* match the experimental data of step 16?

    Magnitude(DB) = $20\log_{10}\{[Z(R)^2 + Z(IMG)^2]^{1/2}\}$ = _____

    Phase(°) = ATAN[Z(IMG)/Z(R)] = _____

18. Based on the results of Figure 4.12, circle all the components listed below that are zero (or near zero) at the resonant frequency: (Is this as expected?)

    Polar magnitude   **Rectangular real**

    Polar phase   **Rectangular imaginary**

The Print and Plot pseudocomponents

Instead of always generating graphs under Probe, sometimes it is more convenient to send results to the output file in the form of tables or plots. For this purpose we can select from the *pseudocomponents* of Table 4.1.

19. As shown in Figure 4.13, add pseudocomponnets *VPRINT1* and *VPLOT1* to your tank circuit and set the attributes as shown. (VPRINT1 and VPLOT1 are from library *special.slb*.)

Symbol	Description
IPLOT	Plot showing current through a cut in the net (Must be placed in series)
IPRINT	Table showing current through a cut in the net
PRINTDGTLCHG	Table showing digital changes during a transient analysis at the connect point
VPLOT1	Plot showing voltages at the connect point
VPLOT2	Plot showing voltage differentials between two connect points
VPRINT1	Table showing voltages at the connect point
VPRINT2	Table showing voltage differentials between two connect points

Table 4.1

Printpoint
pseudocomponents

20. Run PSpice and note the results at the end of the output file. Did you observe a table of frequency versus voltage values and a plot of frequency versus phase values at the *Vtest* node? Were the number of table entries equal to the number of specified points (50)?

    Yes   **No**

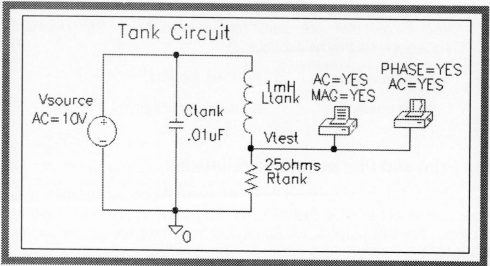

Figure 4.13

VPRINT1 and
VPLOT1 added

Advanced Activities

21. To prove the equivalency of the polar and rectangular formats, we will generalize step 17 by generating complete polar waveforms from rectangular operators.

    Enter the following equations into any appropriate AC sweep Probe graph (separate Y-axes) and note the waveforms produced. (Reminder: ATAN gives results in radians.) Comparing the waveforms to Figure 4.12, what conclusions can you draw?

SQRT(PWR(R(V(Vsource:+)/I(Vsource)),2)+PWR(IMG(V(Vsource:+)/I(Vsource)),2))

ATAN(IMG(V(Vsource:+)/I(Vsource))/R(V(Vsource:+)/I(Vsource)))*180/3.14

22. Using PSpice in the AC sweep mode, plot *power* in both polar and rectangular formats for a high-Q tank circuit (Figure 4.1 with Rtank = 10Ω). Do the results show that most power is apparent power (rather than real power)?

EXERCISES

- In the frequency domain, the simple RC circuit of Chapter 3 is known as a *low-pass filter*. Using PSpice, generate a Bode plot and determine its *break frequency* (V_{out} = .707 × V_{in}) and *rolloff* (slope of high-frequency portion in dB/dec).

QUESTIONS AND PROBLEMS

1. Why is the impedance of a tank circuit the greatest at the resonant frequency?   (<u>Hint</u>: What happens to the current in the two branches near the resonant frequency?)

2. An expression such as V(C1:2)*I(V1) produces what Y-axis unit of measurement?

3. How does Figure 4.12 (rectangular format) prove that the tank circuit is:

   (a) resistive at the resonant frequency?

   (b) capacitive at frequencies above the resonant frequency, and inductive at frequencies below the resonant frequency?

4.   What voltage ratio is represented by +10dB?   -10dB?

5.   List two ways that circuit Q can be defined for a tank circuit.

6.   Why is real power generated in a tank circuit greatest at lower frequencies?
     (<u>Hint</u>: Which branch receives most of the current at low frequencies?)

CHAPTER 5

Families of Curves
The Parametric Mode

OBJECTIVES

- To use the *parametric* mode to plot a family of curves for the tank circuit of Chapter 4.
- To expand or compress waveforms for better viewing.

DISCUSSION

In Chapter 4 we analyzed the tank circuit of Figure 5.1 in the frequency domain, and we generated a Bode plot.

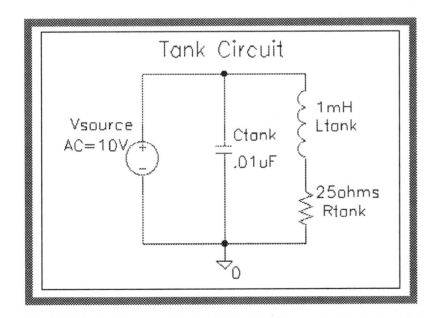

FIGURE 5.1

Tank circuit

However, for many applications, it becomes convenient and informative to show the properties of a tank circuit as a *family* of curves with various values of Q (X_L/Rtank). Such a plot is shown in Figure 5.2, and is the subject of this chapter.

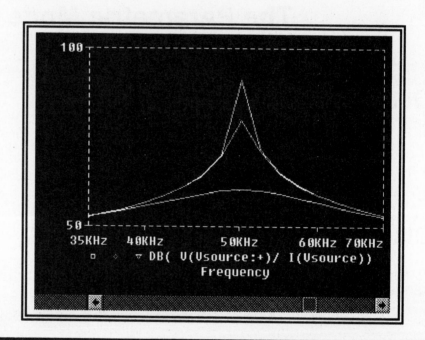

FIGURE 5.2

Family of curves of various values of Q

NESTED CURVES

The plot of Figure 5.2 was created with a *nested sweep* (one sweep inside another). The first (outer) sweep we call the *main* sweep, and the second (inner) sweep we call the *nested* (parametric) sweep.

The main sweep variable is usually related to the X-axis and the nested sweep variable creates the family of curves. In Figure 5.2, the main sweep variable is frequency and the nested sweep variable is Rtank.

THEORY

In this experiment we will sweep Rtank through the values 1, 10, and 100Ω. For a resonant frequency of 50.33kHz, the following equations determine the circuit Qs.

$Q(@1\Omega)$ = X_L (@ f_r) / Rtank = (6.28 x 50.33kH × 1mH) /1 = 316

$Q(@10\Omega)$ = X_L (@ f_r) / Rtank = (6.28 x 50.33kH × 1mH) /10 = 31.6

$Q(@100\Omega)$ = X_L (@ f_r) / Rtank = (6.28 x 50.33kH × 1mH) /100 = 3.16

SIMULATION PRACTICE

Draw the circuit

1. Draw or bring back the tank circuit of Figure 5.1 (from Chapter 4).

Select the sweep mode

2. The *main* sweep variable (frequency) should already be set. (If necessary, review Chapter 4 to sweep Vsource from 10kHz to 100kHz at 50 pts/decade.)

3. To sweep a *global* variable (such as Rtank), we follow the three-step sequence used in Chapter 2 (*Schematics Note 2.2*). However, because Rtank is now a *nested* sweep variable (rather than a *main* sweep variable), we use the *Parametric* dialog box to set up the parameters in step 3.

   Step 1: Change 25ohms to {RVAL}.

   Step 2: Define RVAL (**Draw**, **Get New Part**, **Browse**, **special.slb**, **param**, **OK**, **DRAG** box to any convenient location, **CLICKL**, **CLICKR**. **DCLICKL** on PARAMETERS, Name1 = RVAL, **Save Attr**, Value1 = 25, **Save Attr**, **OK**).

   > Reminder: *Value1 = 25* sets the Rtank value used for an optional DC-only *bias point* analysis.)

   Step 3: Set up the parametric sweep: **Analysis**, **Setup**, **CLICKL** on enabled box for Parametric mode, **Parametric** and enter parameters as shown in Figure 5.3, **OK**, **Close**. (Note: By enabling the *Value List*, we sweep Rtank through the numbers listed in the *Values* box.)

   At the completion of all three steps, your schematic should resemble Figure 5.4.

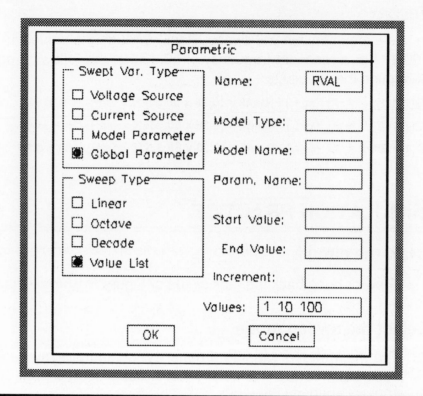

FIGURE 5.3
Parametric
dialog box

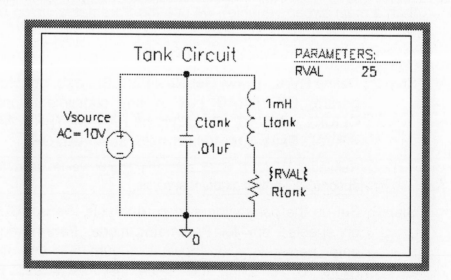

FIGURE 5.4
Tank circuit setup
for nested sweep

Analyze the circuit

4. Begin execution (**Analysis**, **Simulate**) and bring up the PSpice window. Note how the frequency calculations repeat in a *nested* manner for Rtank values of 1, 10, and 100.

    After the calculations are done, the *Available Sections* dialog box appears (Figure 5.5), showing the range of parameter (RVAL) values and giving us an opportunity to select only those we wish to display. Since we wish to display all: **All**, **OK**.

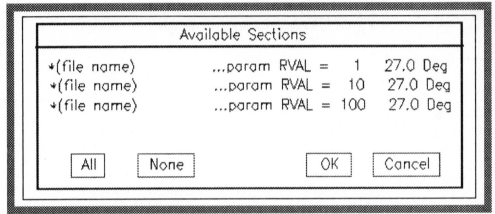

FIGURE 5.5

The Available Sections dialog box

5. The initial Probe graph appears (Figure 5.6) with the main sweep variable displayed by default.

Select the X- and Y-axes variables

6. The X-axis default variable and range are proper. To display a family of impedance curves on the Y-axis (Figure 5.7): **Trace**, **Add**, enter *DB(V(Vsource:+)/ I(Vsource))* into the *Trace Command* box, **OK**.

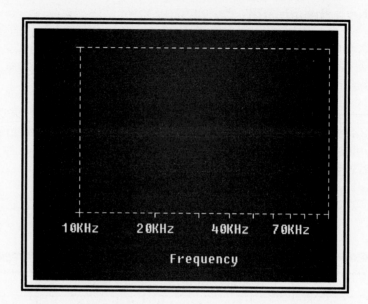

FIGURE 5.6

The initial
Probe graph

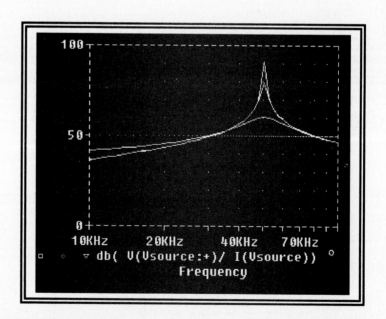

FIGURE 5.7

Family of curves
plot

7. Viewing Figure 5.7, it is clear we have a resolution problem: The "interesting" portion of the curve is too small for easy viewing. After reviewing *Probe Note 5.1,* zoom in on the resonant portion of the curves so your graph resembles Figure 5.2.

Probe Note 5.1
How do I expand or compress waveforms?

We have two techniques for accomplishing this: Change the X- and Y-axes ranges, or zoom in (or out) on portions of the waveform.

- To reset X- and Y-axes ranges: **Plot, X-axis Settings** or **Y-axis Settings, User Defined** (in *Data Range* section), enter the desired starting and ending ranges (such as 35k to 70k), **OK**.

- To zoom in (or out) on portions of a waveform: **View, In** (or **Out**) DRAG crosshairs to desired point, **CLICKL**.   Another more versatile method of zooming in allows us to choose the exact area to expand. To accomplish this: **View, Area,** DRAG crosshairs to any *corner* of desired expansion area, **CLICKLH** and drag to create a box of the desired size, release button.

In all cases, **View, Fit** to bring back the original waveform, **View, Previous** to bring back previous waveforms, and **View, Pan - New Center** to set a new center-of-screen.

When either the X or Y-axis is expanded, scroll bars automatically appear, which enable us to scroll through the normal data range.

---

8. As suggested within *Probe Note 5.1*, use the X-axis scroll bars to position the curves and to examine the data about the resonance portion.

9. Good documentation demands that the waveforms of Figure 5.2 be labeled for clarity. Review *Probe Note 4.3* of the last chapter and label the curves of your graph. (A suggested labeling format is shown in Figure 5.8.)

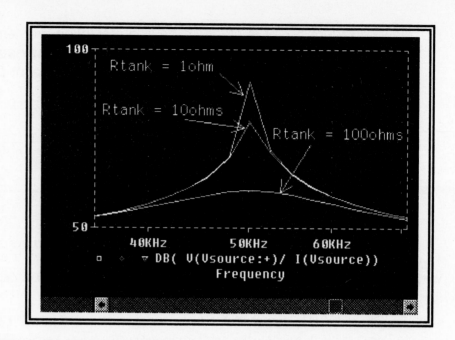

FIGURE 5.8

Final expanded and
documented graph

10. Based on your results (Figure 5.8), use cursors to fill in the resonant frequency and the Q of each plot below (Q = f_r/BW). Compare with the values calculated in the discussion. (<u>Note</u>: Because Q is an approximation, do not except close correlation between theory and experiment.)

f_r =   _____

Q(@1Ω)   =   _____

Q(@10Ω)   =   _____

Q(@100Ω) =   _____

11. To perform an optional bias point analysis, set *DC = 100V* (**DCLICKL** on Vsource's symbol) and run a simulation. Does the SMALL SIGNAL BIAS SOLUTION listed in the output file show a Vsource current of 100A, 10A, and 1A for each of the parametric values of Rtank (1Ω, 10Ω, and 100Ω)?

Yes   **No**

12. To perform an optional bias point analysis using *RVAL = 25*, disable all sweeps (AC and parametric), set *DC = 100V* and run a simulation. Does the SMALL SIGNAL BIAS SOLUTION listed in the output file show a Vsource current of 4A (100V/25Ω)?

        Yes   No

Advanced Activities

13. Using the phase operator P, add a family of phase-angle curves for the circuit impedance to your graph (as shown in Figure 5.9).

14. Using the R (real) and IMG (imaginary) operators, open a second window and plot graphs of the real and imaginary components of the impedance operator. Compare your results to the polar plots of Figure 5.9.

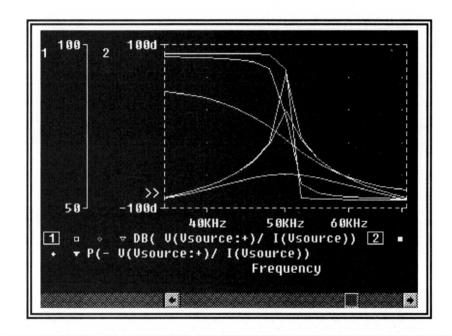

FIGURE 5.9

Adding impedance phase

EXERCISES

- Treating the RC circuit of Chapter 3 as a *low-pass filter*, generate a Bode plot that gives a family of curves for various values of R1.

PSpice for Windows

- With the help of the parametric mode, design a tank circuit suitable for an AM radio station. It must have a resonant frequency of 680kHz and a bandwidth of 6kHz.

QUESTIONS & PROBLEMS

1.   To create a high-Q tank circuit, use a

  (a) low value of Rtank.
  (b) high value of Rtank.

2.   Can a resistor be a main sweep variable?

3.   Which of the following analysis options is used to specify the variable that generates the family-of-curves?:

  (a) Parametric
  (b) AC Sweep

4.   What two general methods are available for expanding (zooming in on) a waveform?

5.   Referring to Figure 5.9, the phase-angle curve with the sharpest changes corresponds to which of the following?:

  (a) Rtank = 1Ω
  (b) Rtank = 10Ω
  (c) Rtank = 100Ω

6.   At resonance, would you expect the real part of the impedance operator to be large or small when compared to the imaginary part?   Why?

CHAPTER 6

The RC Time Constant
Pulsed Waveforms

OBJECTIVES

- To determine the RC and LR response to a square wave.
- To generate multiple plots within a single *Probe* window.

DISCUSSION

When a sine wave powers an AC circuit, it usually means an analog application in the frequency domain. When a *pulsed* waveform powers the same circuit, it usually means a digital application in the time domain.

In this chapter, we will determine the transient response of the simple RC circuit of Figure 6.1 to a pulsed voltage source.

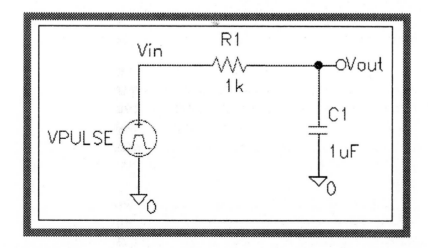

FIGURE 6.1

The pulsed
RC circuit

PSpice for Windows

LABELING NODES

Looking at Figure 6.1, we have often found it difficult to determine the reference designator of the output node. For example, is it V(R1:1) or V(R1:2) or V(C1:1) or V(C1:2)? One solution is to label various wire segments, and a second method is to assign and label a *bubble* (as shown by *Vin* and *Vout* in Figure 6.1).

In this and all future chapters, we will put these new labeling practices to good use.

THEORY

When working in the time domain, Kirchoff's law generates the following differential equation for the circuit of Figure 6.1:

$$V_{in} = V_{pulse} = R \, dQ/dt \; + \; Q/C$$

The following solutions govern the rise and fall times of the RC circuit.   (The term "RC" is known as a *time constant* and in our case has the value 1ms.)

$$V_{out}(rise) = V_C = V_{peak}(1 - e^{-t/RC})$$

$$V_{out}(fall) = V_C = V_{peak} \, e^{-t/RC}$$

SIMULATION PRACTICE

Draw the circuit

1.   Draw the RC test circuit of Figure 6.1 and set the attributes as shown. (Be sure to label the input and output nodes using the techniques of *Schematics Note 6.1*.)

2.   To generate the desired input waveform, program VPULSE as follows: **DCLICKL** on the VPULSE symbol to open up its *Part Name* dialog box, and enter the seven values specified by Figure 6.2. (Display any or all of these attributes if you wish.)

3.   The pulse width is equal to how many RC time constants?

   PW (pulse width) = _____ time constants

Schematics Note 6.1
How do I label wire segments and bubbles?

- To label any wire segment: **DCLICKL** on a wire segment to bring up the *Set Attribute Value* dialog box, enter any desired label (such as Vin), **OK**.

- To assign and label a bubble: **Draw, Get New Part, BROWSE, port.slb, BUBBLE, OK, DRAG** bubble to desired location, **CLICKL, CLICKR**. **DCLICKL** on part BUBBLE to bring up *Set Attribute Value* dialog box, enter *Vout* (or label of your choice), **OK**.

  These wire and bubble attributes turn into trace variables [such as V(Vin) and V (Vout)] and can be used by *Probe* to generate waveforms.

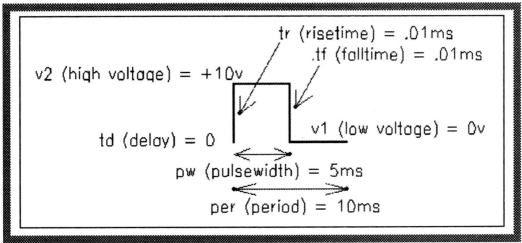

FIGURE 6.2

VPULSE attribute
specifications

Select the sweep mode

4. Bring up the transient dialog box (**Analysis**, **Setup**, etc.) and set up the sweep variable as time from 0 to 10ms. (Remember, the initial time is zero by default.)

Analyze the circuit & set the X- and Y-axes variables

5. Run PSpice (**Analysis**, **Simulate**) and generate the default *Probe* graph. Using our new wire segment trace variables, display the curves of Figure 6.3. [**Trace**, **Add**, **CLICKL** on V(Vin), **CLICKL** on V(Vout), **OK**.]

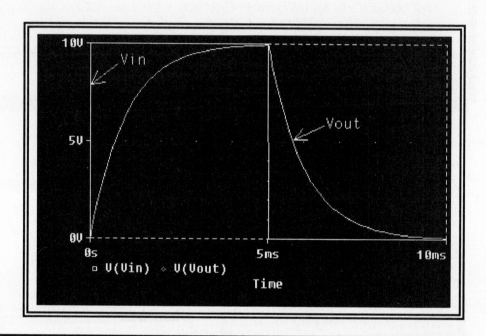

FIGURE 6.3

RC response for
five time constants

6. Using the cursors, carefully measure the rise and fall voltage levels for each of the time constants listed next: (Hint: Refer to *Probe Note 6.1* to generate the marked points of Figure 6.4.)

    1 time constant (1ms)   = _____

    4 time constants (4ms) = _____

    6 time constants (6ms) = _____

    9 time constants (9ms) = _____

Probe Note 6.1
How do I mark coordinate values on my graphs?

To mark the coordinate values of any point on a graph:

1. Activate the cursor (**Tools, Cursor, Display**).
2. Assign each cursor to the desired reference designator, place the cursor at the desired location, and **Tools, Label, Mark**.

   For convenience, these marked points can be repositioned anywhere on the screen. To accomplish this, first disable the cursor (**Tools, Cursor, Display**).   Then, select the desired marked point, **CLICKLH** and drag to new location.   Use the same technique to rotate or lengthen any line segment pointer to a marked point value. To remove a marked point **CLICKL** to select, **Edit, Delete**.

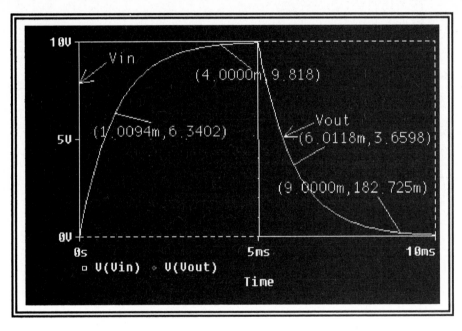

FIGURE 6.4

Showing marked points

7.   Repeat step 6 using a calculator on the theoretical equations developed in the discussion. (Do they match Figure 6.4?)

   1 time constant (1ms) = _____

   4 time constants (4ms) = _____

   6 time constants (6ms) = _____

   9 time constants (9ms) = _____

8. Using the *Mark Voltage Differential* marker as shown in Figure 6.5 [or **Trace**, **Add**, enter V(Vin,Vout), **OK**], add a plot of the voltage *across* R1 and generate the resistor waveforms of Figure 6.6. [Does Vin = V(Vin,Vout) + Vout?]

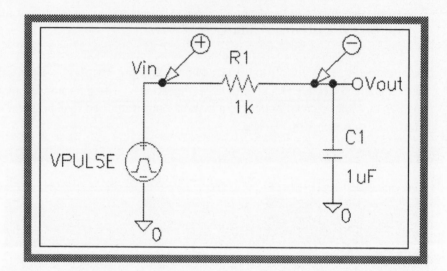

FIGURE 6.5

Marking a
differential voltage

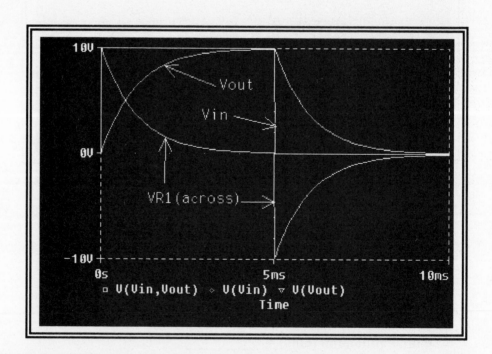

FIGURE 6.6

Vr and Vc waveforms

PSpice for Windows

9.   A possible problem with Figure 6.6 is that there are too many curves on a single graph.   To solve the problem, first clear the *Probe* window (rerun PSpice, or delete the cursor, all marked points, and all curves).   Based on *Probe Note 6.2*, generate the multiple-plot graph of Figure 6.7.

Probe Note 6.2
How do I generate multiple plots within a single Probe window?

To add up to three additional plots to your Probe window (for a total of four): **Plot, Add Plot,** etc. (up to three times).

To select a plot, **CLICKL** anywhere within the desired plot.   Note that "SEL>>" moves to the selected plot.

To add curves to the selected plot, follow the usual sequence (**Markers** or **Trace Add**).

To delete plots, select plot, **Plot, Delete Plot**.

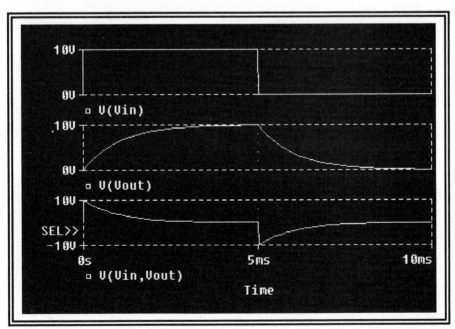

FIGURE 6.7

Display of voltages using multiple plots

PSpice for Windows

10. Using the techniques of Probe Note 6.3, label each of the graphs of Figure 6.7. (A possible result is shown in Figure 6.8.)

Probe Note 6.3
How do I label the Y-axis?

To label any Y-axis, select the desired plot (**CLICKL** within plot), **Plot, Y-Axis Settings** to bring up the "Y-axis Settings" dialog box, enter desired label into "Axis Title" block, **OK**. If a given plot contains more than one Y-axis, enter the appropriate number into the "Y-axis" box.

Any label can be changed by repeating the process.

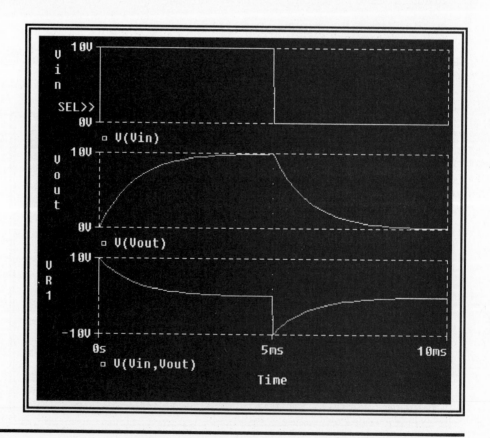

FIGURE 6.8

Adding labels
to graphs

PSpice for Windows

Long and short RC time constants

11. The curves of Figure 6.8 were generated by choosing a value of R1 in which the pulse width (PW) equals five RC time constants (5 × 1ms = 5ms).   This is a *medium* RC time constant value.

Using ratios, complete the following table and determine the values of R1 (C1 remains constant) that will produce the *short* (PW = 1 time constant) and *long* (PW = 25 time constants) conditions indicated:

- **Medium:**   5 × (R1 × 1μF) = 5ms, R1 = ___1kΩ___
- **Long:**   25 × (R1 × 1μF) = 5ms, R1 = _____
- **Short:**   1 × (R1 × 1μF) = 5ms, R1 = _____

12. To show these three cases, sweep R1 as a *nested* (parametric) variable with the values determined in step 11 and generate the graph of Figure 6.9. If necessary, review Chapters 2 and 5, and refer to the three-step process outlined:

- Set R1's <u>value</u> (presently 1k) to *{RVAL}*.
- Define the parameter with **Draw**, **Get New Part**, **Browse**, **special.slb**, **param**, etc.
- Set up the parametric sweep mode with **Analysis**, **Setup**, **Parametric**, etc.

13. Complete the labeling process of Figure 6.9 by placing *short*, *medium*, and *long* on the VR1 plot (the bottom plot).

Advanced Activities

14. Referring to the curves of Figure 6.9, which of the choices below *best* demonstrates each of the following (circle the correct answer)?:

- A *differentiator* in which the output voltage waveform is a measure of the *rate of change* of the input waveform.

    $V(C1)_{long}$ $V(C1)_{short}$ $V(R1)_{long}$ $V(R1)_{short}$

- An *integrator* in which the output voltage waveform is a measure of how the input voltage *accumulates over time*.

    $V(C1)_{long}$ $V(C1)_{short}$ $V(R1)_{long}$ $V(R1)_{short}$

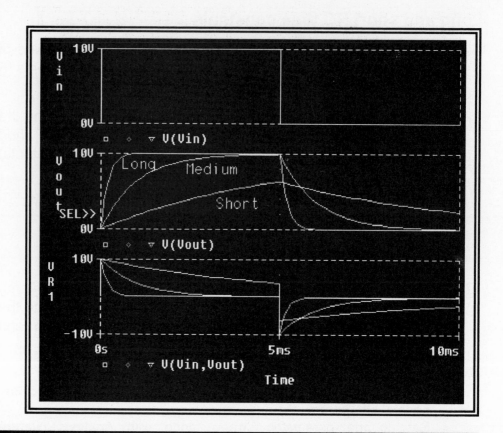

FIGURE 6.9

Long and short
time constants

15. Using the "d" (differentiate) and "s" (integrate) operators, differentiate and integrate VPULSE to generate the curves of Figure 6.10. Compare your results to Figure 6.9. (If necessary, make corrections to your answers of step 14.)

16. By repeating the steps used for the RC circuit, generate a graph for the L/R circuit of Figure 6.11 that is similar to Figure 6.9. (Under what conditions does the L/R circuit approximate differentiator or integrator action?)

17. Perform an energy analysis of the RC circuit of Figure 6.1. (Suggestion: Plot resistive, capacitive, and source energy as separate curves.)

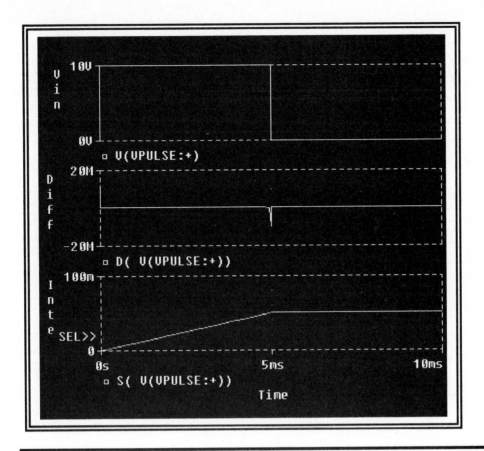

FIGURE 6.10

Using the "d" and "s" math functions

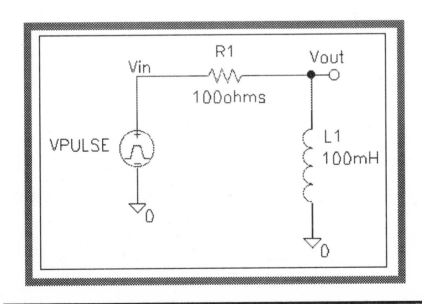

FIGURE 6.11

The L/R circuit

EXERCISES

- Using an RC combination, design a "ones-delay" circuit that delays the logic one 5V pulse to a TTL gate by 5µs. (<u>Note</u>: The logic 1 threshold is 2.73V.)

QUESTIONS & PROBLEMS

1.   When the input to an RC circuit suddenly changes, across what component does most of the change *initially* occur?

   (a) the resistor
   (b) the capacitor

2.   In order for the output from the RC circuit of Figure 6.1 to resemble the input, which of the following would you choose?:

   (a) a very long RC time constant
   (b) a very short RC time constant

3.   Based on the results of Figure 6.10, what is the difference between *integration* and *differentiation*?

4.   In one RC time constant, the voltage rises to _____% of the way from the initial to the final steady-state voltage.

5.   For the Vin waveform of Figure 6.12, sketch directly on the graph the *approximate* Vr and Vc curves (assuming a medium RC time constant).

6.   Referring to the simple RC circuit of Figure 6.1, the break frequency equals $1/(2\pi RC)$ and the rise time (to 63% of maximum) equals RC.   Derive an equation that relates the break frequency to the rise time.

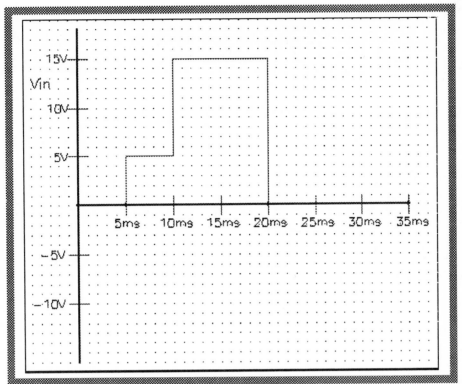

FIGURE 6.12

RC Circuit
waveforms

CHAPTER 7

An Analog Computer
The Stimulus Editor

OBJECTIVES

- To analyze and design an automotive suspension system using an LCR resonant circuit as an analog computer.
- To use the *stimulus editor* to compose and change stimulus signals quickly.
- To review and apply the *Schematic*, *PSpice*, and *Probe* techniques of Chapters 1 through 6.

DISCUSSION

This is the last chapter involving only DC/AC circuits. New material is included, but our primary purpose is to review past chapters and to test your readiness to enter Part II. Therefore, expect to carry out many of the PSpice activities yourself, without the detailed directions of the earlier chapters. (Refer to earlier material as necessary.)

THE AUTOMOTIVE SUSPENSION / LCR CIRCUIT ANALOGY

Our task is to construct an analog computer, one that uses an LCR series circuit to simulate the mass/spring/shock-absorber combination of an automotive suspension system. Such a system exhibits a *damped resonant* response when "hit" with a force.

When designing our suspension system, the major problem will be to choose the correct level of damping. *Underdamping* will cause unwanted oscillations; *overdamping* will make the response too stiff.

Looking at Figure 7.1, we choose a series resonant circuit. This is because in a real suspension system all three components (mass, spring, and shock) are connected at the same point (wheel hub) and therefore have the same velocity. Since velocity is analogous to current, a series circuit is required. Table 7.1 shows the equivalency between the real mechanical system and the electronic simulation system.

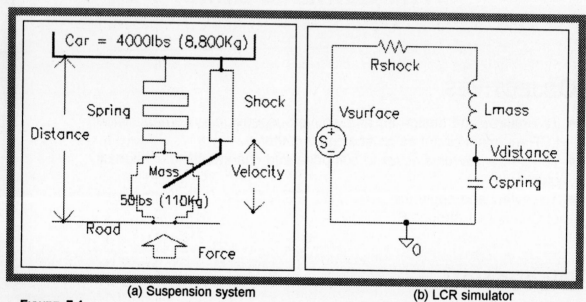

(a) Suspension system   (b) LCR simulator

FIGURE 7.1

Mechanical/simulation
comparison
(a) Suspension system
(b) LCR simulator

Mechanical	Electronic
D (Distance)	Q (Charge)
V (Velocity)	I (Current)
F (Force)	V (Voltage)
M (Mass)	L (Inductance)
K_S (Spring constant)	1/C (Inverse capacitance)
K_f (Coefficient of friction)	R (Resistance)
F (Force on spring) = $K_S D$	V = 1/C x Q
F (Force of friction) = $K_f V$	V = IR (Ohm's law)
F = Mdv/dt (Newton's second law)	V = LdI/dt (Inductor voltage)
$1/2MV^2$ (Kinetic energy)	$1/2LI^2$ (Magnetic energy)
$1/2K_S D^2$ (Potential energy)	$1/2CV^2$ (Capacitive energy)

TABLE 7.1

The mechanical/electronic
equivalency

Once we have chosen the correct spring (C) and mass (L) values, our goal will be to select the correct value of the shock absorber coefficient of friction (R) to provide the desired level of damping.

THEORY

To analyze the system theoretically, we see from Figure 7.1(b) that L represents the tire mass, 1/C represents the spring constant, and R represents the shock coefficient of friction. Because Q (charge) is analogous to distance and proportional to voltage (Q = CV), graphing the capacitor voltage gives a profile of the wheel motion. As designers, our task is to choose appropriate values for R, L, and C.

The calculations are as follows:

- When a 4000lb car (1000lbs/tire) is placed on its springs, they compress approximately 0.5 feet. Therefore K_s = 1000lbs / 0.5ft = 2000lbs/ft. Since M (the mass of a tire) is approximately 2 slugs (64lbs/32ft/sec^2), the ratio of K_s/M is 2000/2 = 1000. It follows that the ratio of 1/C to L (or 1/LC) is also approximately 1000. We choose C = 100μF and L = 10H.

- Because R is our design variable, we will simply choose a low value to begin with—one that represents a very worn set of shocks (such as 10Ω).

- Based on the calculated values for L and C, the resonant frequency of the system is:

$$f_r = 1 / [2\pi(LC)^{1/2}] = 1/[2 \times 3.14 \times (10H \times 100\mu F)^{1/2}] \cong 5Hz$$

STIMULUS EDITOR

When analyzing a suspension system, it would be most convenient to be able to change and modify the "road surface" conditions quickly and easily in order to test for a variety of road hazards.

For a transient analysis, this is achieved by using the *Stimulus Editor* feature of PSpice. Voltage source VSTIM is placed on the schematic, and from this *single* source, any number of stimulus options can be created and saved. Then, during testing, any of these options can be quickly and easily selected at any time.

In the chapters to follow, we will make only occasional use of VSTIM and ISTIM. However, whenever a transient analysis is required, feel free to substitute VSTIM or ISTIM for any of the "standard" sources.

SIMULATION PRACTICE

1. Draw the circuit of Figure 7.1(b), using part *VSTIM* (from *source.slb*) for the voltage source. Change the part reference name to *Vsurface*, set the R, L, and C components to the values calculated in the discussion, and be sure to label the output wire segment *Vdistance*.

2. Referring to *Schematics Note 7.1* as necessary, program VSTIM (Vsurface) to match the "worst case" scenario of Figure 7.2—driving over railroad ties at a speed harmonically related to the system's resonant frequency. (The 1s period equals 1/5th the 5Hz resonant frequency of the suspension system.)

Schematics Note 7.1
How do I use the stimulus editor to program VSTIM?

For example, to create the railroad tie waveform of Figure 7.2 we perform the following:

1. **DCLICKL** on part VSTIM to bring up the *Edit Stimulus* dialog box, enter *Vrailroad* (to specify the railroad tie profile as the first test surface), **OK** to bring up the *Stimulus Editor* and *New Stimulus* dialog box.

2. **PULSE** to select a pulse stimulus, **OK** to bring up *Modify Stimulus* dialog box and fill in as specified by Figure 7.2, **Apply**.   Note the waveform displayed. Does it match your specifications? If so, **OK**.

3. When done observing the waveform, **File, Save**. The voltage source is now programmed for the railroad tie surface. (The Stimulus Editor can remain active, but if desired, **File, Exit**.)

---

3. Bring back the Schematics window and note the *Vrailroad* attribute to part *Vsurface*.

      The circuit now complete, set up the system for a transient mode sweep from 0 to 2 seconds (to display approximately 2 cycles).

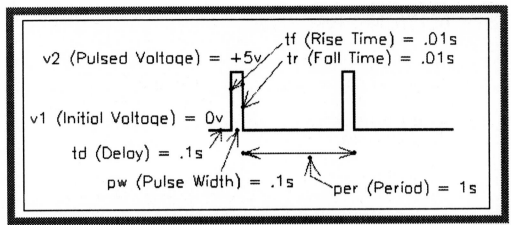

FIGURE 7.2

VPULSE attributes

4. Run PSpice and create the initial Probe graph. (Does the X-axis range from 0 to 2s?)

5. Add traces Vsurface and Vdistance and create the motion graph of Figure 7.3.

6. As we suspected, the waveforms of Figure 7.3 make it clear that the wheel oscillates way too strongly. Circle the correct response below:

    • **The system is *underdamped*.**

    • **The system is *overdamped*.**

7. Based on step 6, we conclude that the value of Rshock is too small. Change Rshock to 1kΩ, reevaluate (run PSpice again), and generate the new waveforms of Figure 7.4. (<u>Note</u>: To create Figure 7.4, use **Plot**, **Y-axis Settings**, etc., to change the default Y-axis range to match that of Figure 7.4.)

8. Viewing the new waveforms (Figure 7.4), it is clear that we went too far the other way. The shock absorber is now too stiff (Rshock is too large) and the system is *overdamped*.

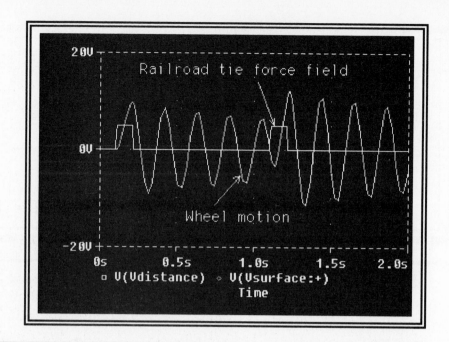

FIGURE 7.3

Response for
Rshock = 10Ω

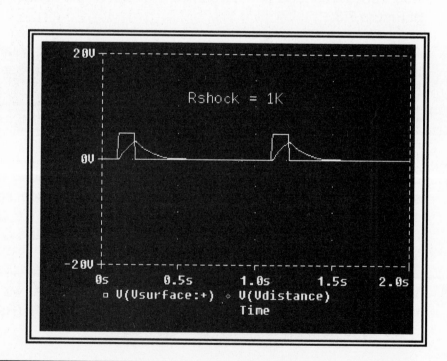

FIGURE 7.4

Response for
Rshock = 1kΩ

PSpice for Windows

Adding a nested sweep variable

9. Clearly, there is an intermediate value between 10Ω and 1kΩ that is just right. To help find the right value, make Rshock a parametric (nested) variable and sweep its value over the following range: Rshock = 50, 100, 300, 500, and 800. The resulting waveform set is shown in Figure 7.5.

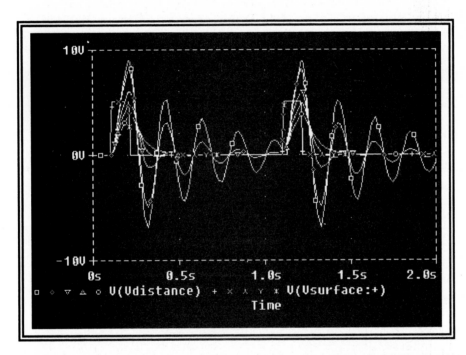

FIGURE 7.5

Waveform set

10. From the waveform set of Figure 7.5, select the one curve that you believe to be the best. (The best system absorbs the greatest amount of shock, but without *excessive* oscillation.) What value of Rshock did you choose?

      Rshock (best curve) = _____

11. We wish to display this one "chosen" curve, but do not wish to repeat the analysis process. Delete the present curves, and review *Probe Note 7.1* to display the single "best" curve (such as shown in Figure 7.6).

PSpice for Windows

Probe Note 7.1
How do I single out individual curves from a family of curves?

To single out individual curves, we use the "at" symbol (@). For example, the graph of Figure 7.5 contains five curves in each group, numbered from 1 to 5.   Curve 1 corresponds to R1 = 50Ω and curve 5 corresponds to Rshock = 800Ω.   For our "best" curve, we choose Rshock = 300Ω.

- To draw curve 3 of Vsurface: **Trace, Add Trace**, enter V(Vsurface:+)@3, **OK**.
- To draw curve 3 of Vdistance: **Trace, Add Trace**, enter V(Vdistance)@3, **OK**.

The result of these two operations is shown in Figure 7.6.

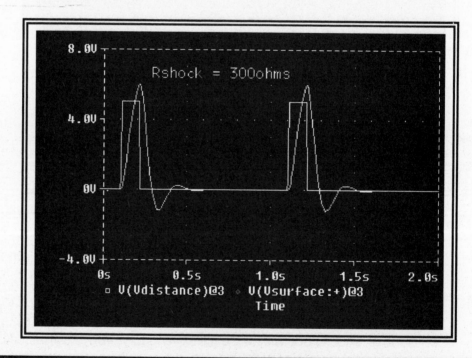

FIGURE 7.6

Displaying individual
curves

Adding new stimuli to Vsurface

Having successfully designed our suspension system for a series of railroad ties, we now wish to test our design on another hazard. We choose the single pothole surface hazard of Figure 7.7. Because the waveform is "one-time-only" (nonrepetitive), we will choose the PWL (piecewise linear) mode.

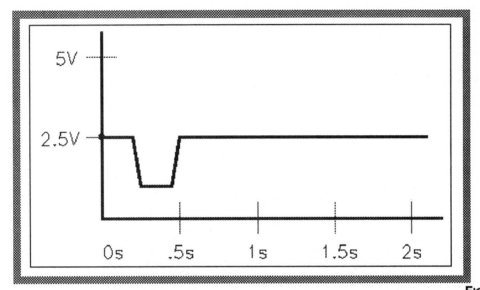

FIGURE 7.7

Pothole road surface hazard

12. Return to the Schematics window, change Rshock from {RVAL} to the "best" value (300Ω?), and disable the parametric analysis.

13. Return to the Stimulus Editor (if necessary, **DCLICKL** on part VSTIM). Delete the *Vrailroad* waveform (select the Vrailroad reference designator, **Edit**, **Delete**).

14. If necessary, reset the X- and Y-axis to 2s and 5V, respectively (**Plot**, **Axis Settings**, etc.)

15. **Stimulus**, **New** to bring up the *New Stimulus* dialog box, enter Name *Vpothole*, **PWL**, **OK**. Note that a "pencil" automatically appears.

PSpice for Windows

16. Our first step is to draw an <u>approximate</u> version of the desired waveform. (During this step, it is necessary only to enter the corner points at their approximate positions.)

   With the pencil showing, **CLICKL** at the *approximate* start of the waveform (x ≅ 0 and Y ≅ 2.5V) to place the first corner point—marked by a small box—and create a line from the origin to the first corner point.

   Continue in like manner (**DRAG**, **CLICKL**), until the curve *approximates* that of Figure 7.7.

17. Since your curve is probably not shaped exactly as you wish, reposition the corners as follows: Exit the "pencil" mode (**QLICKR**), **CLICKL** on any corner box to select, **CLICKLH** and drag corner as necessary. (To reenter the "pencil" mode if necessary: **Edit**, **Add Point**.)

> <u>Note</u>: If we place the first corner box (x ≅ 0, y ≅ 2.5V) near the Y-axis, the system will properly ignore the initial step from the origin to this point.

18. **File**, **Save** to save settings to Vpothole. (If you wish, **File**, **Exit** to exit the Stimulus Editor—however, it can remain active.)

19. Bring back the Schematics Window, **DCLICKL** on *attribute Vrailroad*, change to *Vpothole*, **OK**.

20. Run PSpice and generate the curves of Figure 7.8.

21. Viewing the results, does the suspension system appear to handle the pothole as well as it did the railroad ties?

   Yes   No

Advanced Activities

22. Change the speed (width and period) of the "car" as it negotiates the railroad and pothole hazards and report on the results.

23. Referring to *Schematics Note 7.2* as necessary, test your suspension design on a new or modified road surface of your choosing. Sketch the road hazard and the resulting motion on Figure 7.9.

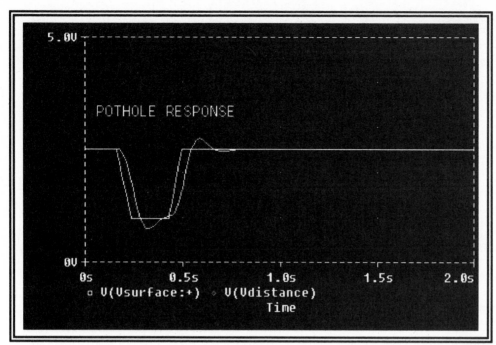

FIGURE 7.8

Pothole test
results

Schematics Note 7.2
How do I add new stimuli, modify old stimuli, and switch between stimuli?

- To add new stimuli:

  DCLICKL on part VSTIM, **Stimulus**, **New**, enter new Name, select Type, enter parameters, **OK**, change X- and Y-axes as necessary, **File**, **Save**.   Bring back the Schematics window, **DCLICKL** on old stimulus name and enter new, **OK**.

- To modify old stimuli:

  DCLICKL on part VSTIM, **Stimulus**, **Get**, select Name, **OK**, **CLICKL** on reference designator to select, **Edit**, **Object**, modify stimulus as desired, **OK**, **File**, **Save**.

- To select any stimuli:

  DCLICKL on part VSTIM name (Vrail, Vpothole, etc.) and enter desired name, **OK**.

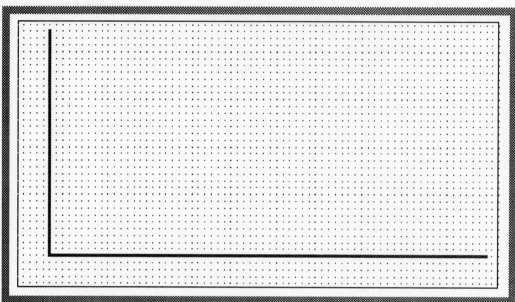

FIGURE 7.9

"Student" road
hazard study

24. To the "best" response curve of Figure 7.6 (or 7.8), add a graph of "best" wheel velocity (Lmass current) to generate a plot like that of Figure 7.10. (Note: The X-axis has been expanded for easier viewing.)

   Examining the results, do you see any relationship between the distance and velocity curves? (Hint: Velocity = slope of distance.) If you wish, use the horizontal scroll bar to scan through the waveform (see *Probe Note 5.1*).

25 Using the "d" (differentiate) operator, generate the velocity (speed) curve by differentiating the distance curve. Is the resulting curve equal to the velocity curve generated by step 24?

EXERCISES

● Do a frequency analysis of the suspension system computer of this chapter and generate a family of Bode plots. How does the Q and bandwidth of your "best" system compare to the underdamped and overdamped cases?

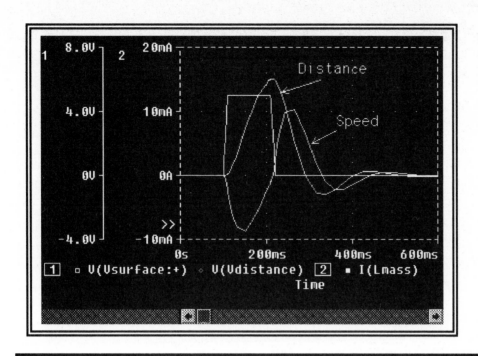

FIGURE 7.10

Adding a plot of "velocity" (current)

- Do an energy study of the suspension system. (Add waveforms of *kinetic energy* of the wheel $(1/2\ LI^2)$, *potential energy* of the spring $(1/2CV^2)$, and *heat energy* of the shock absorber $(1/2\ I^2R)$.

  What does your result say about *conservation of energy*? Does it appear that properly functioning shock absorbers must dissipate a great deal of energy? (Hint: Use the AVG operator to generate a running average of energy.)

QUESTIONS & PROBLEMS

1. What is the difference between an *analog* computer and a *digital* computer?

2. Based on the analogy of this experiment, how does an inductor simulate Newton's first law (An object in motion tends to stay in motion, and an object at rest tends to remain at rest—unless acted on by an external force)?

3. When shocks are worn, the system is

    (a) underdamped
    (b) overdamped

4. Why does a *series* LRC circuit (rather than a *parallel* tank circuit) simulate the suspension system? (<u>Hint</u>: Do all components of the suspension system have the same instantaneous velocity?)

5. Reviewing Figure 7.4, why is the second "bounce" of the wheel greater than the first?

6. Why is driving over railroad ties at a "harmonic" speed a worst case condition?

7. Why could the activities of this chapter be called a *double simulation*?

PART II

Diode Circuits

In Part II, we move from the simple passive **DC** and **AC** components to the active devices of solid-state electronics.

Because we assume that the major **PSpice** techniques of Part I are second nature, we condense the more detailed four-step sequence into a two-step "process summary":

1. What sweep mode will I use?
2. What variables will I assign to the X- and Y-axes?

CHAPTER 8

The Diode
Diode Curves

OBJECTIVES

- To plot diode curves.
- To design a zener diode voltage regulator.
- To determine how the characteristics of a diode are affected by temperature.
- To examine and modify a diode's model.

DISCUSSION

As shown in Figure 8.1, a diode is an active device with a single PN junction.

Using the "waterfall" analogy, electrons in the high-energy conduction band can "fall" easily to the holes in the low-energy valence band when passing from the N to the P region (forward bias voltage). However, electrons cannot easily flow "uphill" from the P to the N region (reverse bias voltage). Therefore, the PN junction offers a low resistance in the forward direction and a high resistance in the reverse direction.

Note that the *schematic symbol* arrow points in the forward bias direction for *conventional* current flow (Figure 8.1).

If enough voltage is applied in the reverse direction, a conventional diode will break down, and chain-reaction ionization (avalanching) will destroy the diode. A *zener diode*, on the other hand, avoids avalanche and is designed to work in the reverse breakdown region as a *voltage source* (or *voltage regulator*).

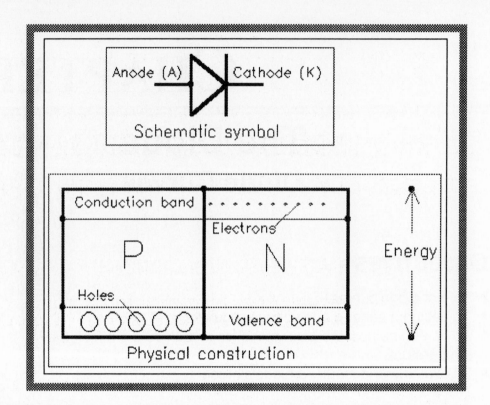

FIGURE 8.1

The PN junction

The best way to display the characteristics of a diode is through the *diode curve*, in which diode voltage and current are plotted on the X- and Y-axes.   In this chapter, we draw diode curves for both the small signal diode and the zener diode.

DEVICE MODELS

As we learned in Chapter 2, all PSpice devices use mathematical *models* and model *parameters* to determine their characteristics.   In the case of the simpler devices used so far (parts R, C, and L), the model parameters are few in number and are only accessible to the designer as attributes.   Furthermore, new parameters cannot be added to their models.

In contrast, parts such as diodes have more complex models, many accessible parameters, and the need to add or modify parameters as needed.   For this reason, MicroSim Corporation placed their model parameters in separate files and a separate directory.

In the case of the 1N4148 and 1N750 diodes, the model parameters are contained in *global* file *eval.lib* in directory *msimev60\lib*. (A global file is accessible to all schematics.)

<u>The local library</u>

When a model library is accessed for the first time (by way of the *Model Editor* dialog box), it is assumed that changes may be made. Therefore, to "protect" the original *global* model, a copy (with any changes made) is automatically written to a *local* model library when the dialog box is exited. (A *local* library is accessible only to the corresponding schematic.)

The new local library is automatically given the same name as the schematic (with extension *.lib*) and is placed in the same directory as the schematic. When the same part model is accessed again, it is the local library that automatically appears in the dialog box. To make sure that the user is aware of its local status, an "X" is placed after the global model name. For even greater flexibility, the user is free to change the local model name at any time.

MODEL PARAMETERS

For the 1N4148 diode of Figure 8.1, its model parameters are listed next:

Parameter	Description	Value
Is	Saturation current	0.1pA
Rs	Parasitic resistance	16Ω
CJO	PN capacitance	2pF
Tt	Transit time	12ns
Bv	Breakdown voltage	100V
ibv	Reverse knee current	0.1A

In this chapter, we will access these model parameters and modify the value of the reverse breakdown voltage (BV).

SIMULATION PRACTICE

The small signal diode

1.   Draw the test circuit of Figure 8.2, and set the attributes as shown.

   • The D1N4148 small signal diode is found in library **eval.slb**. (The D1N4148 symbol that appears on the schematic is its part name.   The model name is also D1N4148,   but does not appear on the schematic.)

   • Because V1 will be swept, and we don't require a bias point solution, its *DC=* voltage value need not be specified.

   • As shown, be sure to label the appropriate wire segment *Vanode.*

   • If you wish, title your schematic as shown.

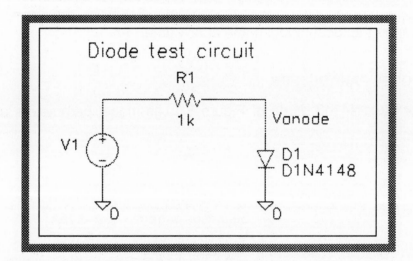

FIGURE 8.2

Diode test circuit

The diode curve

2.   To draw a diode curve, select V1 as the main sweep variable and set up a linear DC Sweep from -110V to +10V in increments of .1V.

3.   Run PSpice, and generate the initial Probe graph of Figure 8.3.

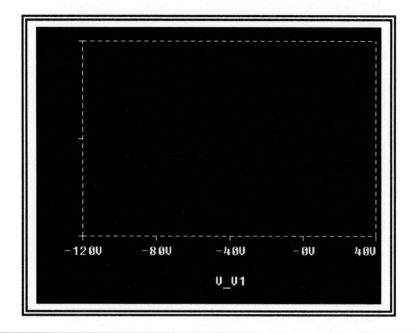

FIGURE 8.3
Initial Probe graph

4. This time, looking at Figure 8.3, the default X-axis variable (V_V1) is not what we need—instead, the X-axis must be the diode's anode-to-cathode voltage [V(Vanode)]. Review *Probe Note 8.1* and make the change.

Probe Note 8.1
How do I change the X-axis variable?

To change the X-axis to any available variable: **Plot, X Axis Settings, Axis Variable, CLICKL** on any listed variable (or enter any allowed value or expression in the "X Axis Variable:" box), **OK, OK**.   (Be aware that the X-axis *range* will automatically be adjusted to match the data base.)

5. Set a current marker at the diode's anode pin [or **Trace**, **Add**, **I(D1)**, **OK**] and recreate the diode curve of Figure 8.4.

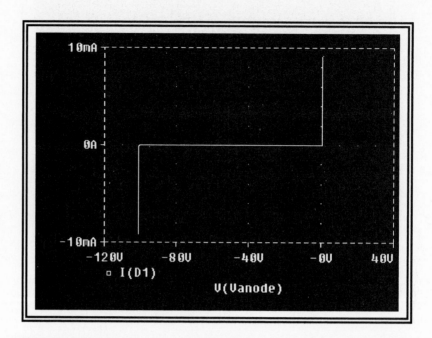

FIGURE 8.4

The diode curve

6. Based on the results of Figure 8.4, determine the following:

    (a) *breakdown voltage* (V_{RSM}) (V_{RSM} is also known as the *non-repetitive peak reverse voltage*. We will arbitrarily measure V_{RSM} @ –1mA.):

    V_{RSM} (@ –1mA) = _____

    (b) *maximum reverse current* (I_R) (I_R is measured at the rated voltage of -100V):

    I_R (@ –100V) = _____

7. When the diode curve is viewed "from a distance" (as in Figure 8.4) it appears to exhibit *ideal* characteristics. That is, a short in the forward direction, and (before breakdown) an open in the reverse direction.

    To bring out the details in the critical forward direction, zoom in on the curve by adjusting the X-axis range (0 to +1V) and generate the graph of Figure 8.5.   (Reminder: Select the **View** menu.)

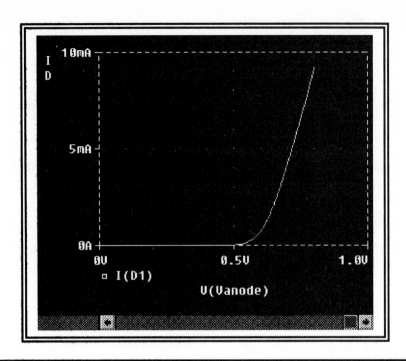

FIGURE 8.5

Forward bias
details

8. From your graph, determine *approximate* values for the following:
    (Hint: Use the cursors.)

    (a) The knee voltage (@ 1mA): _____

    (b) The forward resistance of the diode at 1mA: _____

    [Hint: Place cursor A1 just below 1mA, A2 just above 1mA and
    determine 1/slope as $\Delta V / \Delta I$, **or** use the "d" (differentiate)
    operator to plot 1/dI(D1) from .6V to 1V.]

    > Note: Calculus tells us that $\Delta V / \Delta I$ in the limit as ΔI approaches zero
    > equals dV/dI.   When using the "d" operator, d(Y-axis) performs
    > d(Y-axis)/d(X-axis). Therefore, dI(D1) = dI(D1)/dV(Vanode) and
    > 1/dI(D1) = dV(Vanode)/dI(D1) = forward resistance.

    (c) The forward resistance of the diode at 5mA: _____

Temperature effects

9. All active devices (such as diodes) are sensitive to temperature.
    The diode curve of Figure 8.5 assumes a default temperature of
    27°C (80.6°F). To see how temperature can be changed, review
    Schematics Note 8.1.

Schematics Note 8.1
How do I change the circuit's temperature parameter?

To change the temperature of your circuit: From the *Schematics* window, **Analysis, Setup, CLICKL** on "Enabled" temperature box, **Temperature** to open up the *Temperature Analysis* dialog box, enter the new temperature, **OK, Close.**

---

10. Following *Schematics Note 8.1,* change the temperature of the diode test circuit from 27° (80.6°F) to 60° (140°F).

11. Generate a new diode curve for 60°, compare to steps 6 and 8, and fill in the blanks below.

V_{knee} (@ 1mA, 60°)   = ———   **% change (from 27°)**   = ———

$I_{reverse}$ (@ -100V, 60°)   = ———   **% change (from 27°)**   = ———

DC Sweep nesting

12. A better method of evaluating temperature effects is to generate a family of curves. Following *Schematics Note 8.2,* set up the diode temperature as a nested variable.

Schematics Note 8.2
How do I set up the DC Sweep nested mode?

When adding a nested sweep to the DC Sweep mode, we use the DC sweep *nested* mode (instead of the *parametric* mode that is used with the *AC Sweep* and *Transient* modes) .

As an example of the nested mode, we set up the nested diode curves of step 12 as follows: **Analysis, Setup, DC Sweep** to bring up the (Main) DC Sweep dialog box and fill in as shown in Figure 8.6(a);   **Nested Sweep** to bring up the *DC Nested Sweep* dialog box and fill in as shown in Figure 8.6(b).   **Enable Nested Sweep, OK, Close.**

For future reference, note the following:

- **CLICKL** on the "Nested Sweep" or "Main Sweep" buttons to toggle between boxes.

- Be aware that when the nested sweep variable is a component (such as a resistor) we must follow the three-step "parametric" process outlined in Schematics Note 2.2.

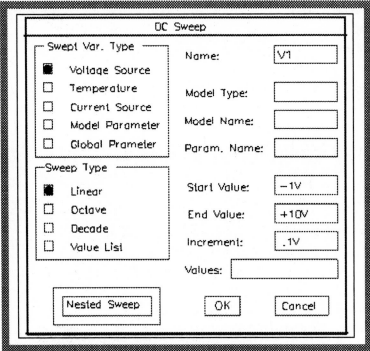

(a) The Main dialog box

(b) The Nested Sweep

FIGURE 8.6

DC Sweep dialog boxes
(a) The Main dialog box
(b) The Nested Sweep

13. Disable the temperature analysis because it interferes with a temperature sweep (**Analysis**, **Setup**, **CLICKL** on "Enabled" temperature box to clear).

14. Run PSpice and create the curves of Figure 8.7.

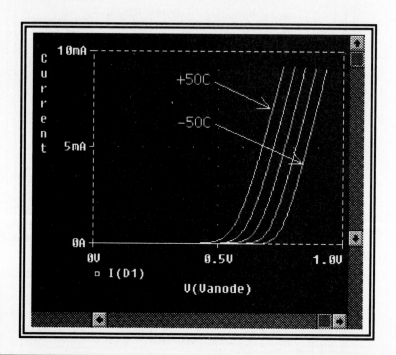

FIGURE 8.7

Family of diode curves

15. Based on the results:

   (a) By what *approximate* percentage does the knee voltage (@ 1mA) change for each degree C?

   % change in knee voltage/°C = _____

   (b) Based on simple inspection of the curves, do the impedance characteristics (1/slope) depend on temperature?

         Yes   **No**

Model parameters

16. Referring to *Schematics Note 8.3*, view the model parameters of the 1N4148 and change the reverse breakdown voltage (Bv) from 100 to 150.

Schematics Note 8.3
How do I examine and change model parameters?

To change a model parameter: Select the component (**CLICKL** to turn red), **Edit**, **Model**, **Edit Instance Model**, examine or change any model parameter, **OK**.

Note that the diode model is a *local* model (did you see the "X" in the model name?).

---

17. Generate a graph similar to that of Figure 8.4. Does the breakdown now occur at -150V?

    Yes   **No**

18. Return the breakdown voltage to -100V (for the next user).

The zener diode

19. Draw the voltage-regulator zener diode circuit of Figure 8.8. (The 1N750 is a 4.7V zener.) Be sure to label the appropriate wire segment *Vout*, as shown.

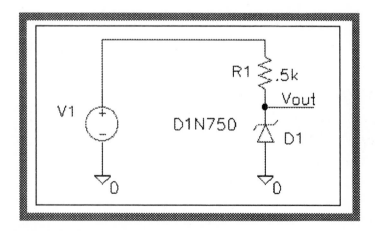

FIGURE 8.8

Zener diode
voltage regulator

20. Generate the Zener graph of Figure 8.9. (If necessary, refer to the *process summary* that follows:)

Listed here is our first *Process Summary*, which condenses the creation of the Zener graph of Figure 8.9 into the following two essential steps: (1) Specify the sweep mode (**Analysis**, **Setup**) and (2) select the X- and Y-axes variables (**Analysis**, **Simulate**).

Here and in the future, you should try to generate all curves on your own, referring to these process summaries only as necessary.

Process Summary for Zener Curve

- The <u>main sweep</u> variable is V1, generated by a DC Main Sweep from 0V to +20V in steps of .1V. (The nested sweep is disabled.)

- The X-axis variable is the negative zener voltage [-V(Vout)], and the Y-axis variable is the zener current [I(D1)].

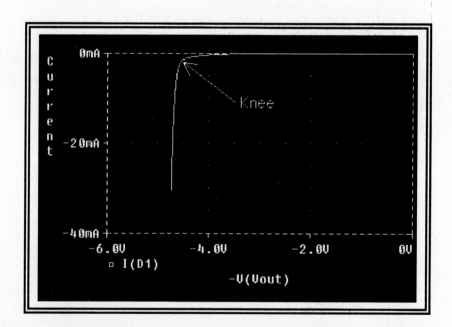

FIGURE 8.9

1N750 zener curve

21. Use zoom techniques and the cursor to determine *accurately* the zener current at a zener voltage of exactly 4.7V. Compare your answer with the 1N750 spec sheet in Appendix D.

I_Z (@ 4.7V) = _____ I_Z (spec sheet) = _____

22. By using the cursor method (see step 8b), determine the zener impedance at the 4.7V test condition and compare to the spec sheet:

    Z(@4.7V) = _____   Z (spec sheet) = _____

23. In the breakdown region, is the zener a *voltage source*? (Is the voltage approximately independent of the current?)

    Yes   No

Advanced Activities

24. Using the "/" (divide) and "d" (differentiation) operators, add a plot of zener impedance to the zener curve of Figure 8.9 (as shown by the expanded curves of Figure 8.10). Do the results agree with step 22 (the PSpice result)?

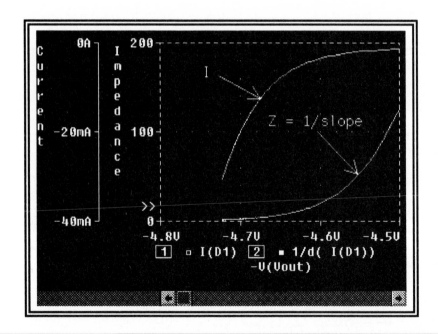

FIGURE 8.10

Adding zener impedance

25. By way of its model parameters, change the reverse breakdown voltage of the 1N750 zener diode from 4.7V to 10V and test the results.

26. Using PSpice, devise a means of determining the capacitance of a small-signal diode when reverse-biased at –1V. (<u>Hint</u>: Consider use of the transient mode and the equation $I = CdV/dt$.)

EXERCISES

- Test the voltage regulation characteristics of a zener diode under a varying load. [<u>Hint</u>: Draw the circuit of Figure 8.11(a) and generate the curve of Figure 8.11(b).] Summarize your results.

QUESTIONS & PROBLEMS

1. The forward-biased AC resistance of a diode is lowest at

    (a) low currents
    (b) high currents

2. As the temperature goes up, the barrier potential goes

    (a) up
    (b) down

3. Using an ohmmeter, give a quick and simple method of checking a diode.

4. What is the difference between a small-signal diode and an *LED*?

5. Based on the results of step 6b, how long would it take 1 coulomb of charge to pass through a reverse-biased diode?

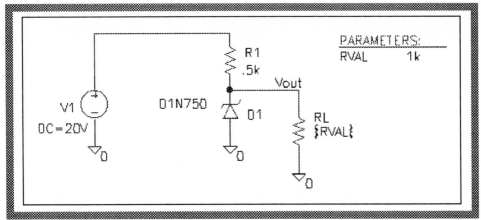

(a) Adding a variable load

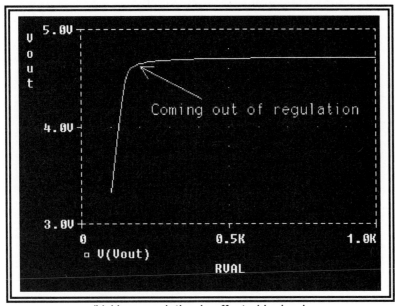

(b) How regulation is affected by load

FIGURE 8.11

Zener regulation
test circuit
(a) Adding a
    variable load
(b) How regulation
    is affected by load

6. A zener diode normally operates

    (a) in forward bias
    (b) in reverse bias

PSpice for Windows

7. Looking to Figure 8.11(b), approximately what is the smallest load value that will maintain zener regulation?

CHAPTER 9

The Power Supply
Voltage Regulation

OBJECTIVES

- To design a power supply, complete with filter and regulator.
- To measure output ripple and regulation efficiency.

DISCUSSION

A "perfect" power supply provides a constant desired voltage, regardless of the value of the load. A "real" power supply, on the other hand, has a ripple and generates an output voltage that varies with the load.

Our goal in this chapter is to design a power supply that approaches "perfection," but without exceeding reasonable size, cost, and complexity limits. Our design philosophy will be to start simple, and to add components and circuits gradually until we achieve our goal.

A REAL VOLTAGE SOURCE

For safety reasons, it is common practice in the laboratory to use a signal generator to simulate the output from a wall socket and step-down transformer. The problem is that such laboratory voltage sources usually have a significant output impedance value. However, the voltage sources used by PSpice are "perfect" and have zero output impedance.

To simulate a laboratory voltage source using PSpice, we have the option of adding a resistor in series with the voltage source, as shown in Figure 9.1. Commonly found values of Zout are 50Ω and 600Ω.

PSpice for Windows

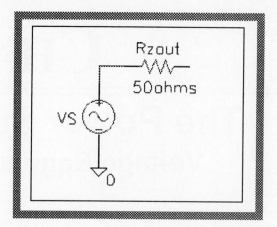

FIGURE 9.1

Simulating Zout of
a power source

Therefore, in this chapter the AC voltage source includes a 50Ω Zout resistor. In future chapters this resistor is optional, but should be added whenever a PSpice simulation is to be directly compared with the same circuit built and tested in the laboratory.

SIMULATION PRACTICE

1. Draw the initial design of Figure 9.2, which is known as a *half-wave rectifier*. (Be sure to label the output wire segment *Vout*, as shown.)

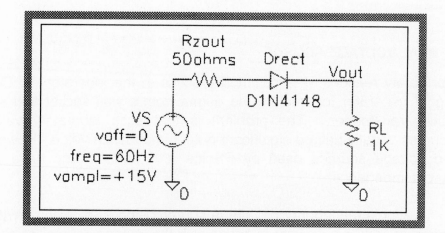

FIGURE 9.2

Half-wave rectifier

2. Using PSpice, generate 5 cycles (10 half cycles) of the output waveform. (We use 10 cycles to better assess steady-state conditions.) Sketch your waveform on Figure 9.3 and label as "*Half-wave rectifier.*" (If necessary, refer to the *Process Summary*.)

> **Process Summary for Half-Wave Rectifier**
>
> - The main sweep variable is time, generated by a transient sweep from 0 to 83ms (10 × 1/120 = 83ms).
>
> - The X-axis is time (by main sweep default), and the Y-axis is Vout [V(Vout)].

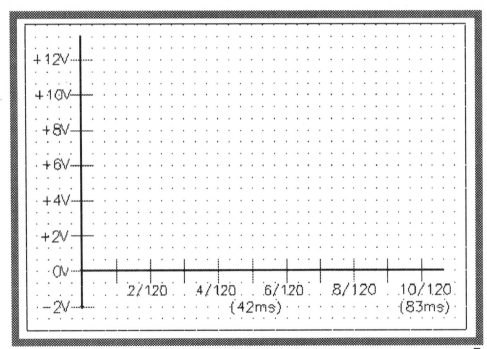

FIGURE 9.3

Power supply output waveforms

Peak rectifier

3. Figure 9.3 clearly tells us that we are a long way from our goal. Let's improve the circuit in two ways:

   - Use a *full-wave rectifier* so both the positive and negative input cycles will power the output.

   - Add a *filter capacitor* to smooth out the waveform.

Make these changes and create the circuit of Figure 9.4, which is known as a *full-wave peak rectifier*. The output voltage will go up and down as capacitor CP charges and discharges. This voltage fluctuation is known as the *ripple* voltage.

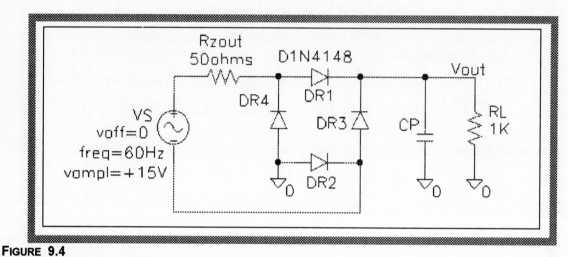

FIGURE 9.4

Full-wave peak
rectifier

4. To complete the design, what value will we choose for CP? A highly accurate answer would require some advanced mathematics. Fortunately, we are only after a "ballpark" estimate and can make some assumptions:

- We assume that a 1V output ripple is acceptable.
- We assume that Vout (average) = 13V (after two diode drops).
- We assume that each capacitive discharge lasts for no more than 1/120 sec.

Using these assumptions on the differential form of $Q = CV$ ($I = C\Delta V/\Delta t$), solve the equation below and determine the approximate value of C:

I (discharge current) = CΔV / Δt =   Vout / RL, where

ΔV	=	ripple voltage	=	1V
Δt	=	discharge time	=	1/120 sec
RL	=	load	=	1kΩ
Vout	=	output voltage	=	+13V

Solving for C yields approximately _____

5. Round off the value of C determined in step 4 (100μF?) and assign to CP. Generate the output waveform, and add your new output voltage curve to Figure 9.3. Be sure to label the curve "*Peak rectifier*."

6. Determine the ripple voltage at the far right of the curve as it approaches steady-state conditions.

    (Suggestion: Review *Probe Note 9.1* and perform the following sequence: Position cursor 1, **Peak**, position cursor 2, **Trough**, report the Y-axis difference.)

    V_{ripple} = _____

    Is the ripple less than 1V, as predicted by our previous calculations?

    Yes   **No**

Probe Note 9.1
Does Probe include any "tricks" to help measure ripple and other waveform parameters?

The list below includes some rather clever measurement methods possible under the Probe "cursor" menu.   In all cases, the cursor affected (1 or 2) is the last one assigned to a waveform or moved.

First, enable the cursor and bring up the cursor window by **Tools, Cursor, Display**. Then, **Tools, Cursor**, followed by any of the following:

- **Peak** moves the cursor in the same direction as its last movement and finds the first peak (a data point with lower values on each side). Repeat, and the cursor will find the next peak.
- **Trough** moves the cursor in the same direction as its last movement and finds the first trough (a data point with higher values on each side).
- **Slope** moves the cursor in the same direction as its last movement and finds the next maximum slope (positive or negative).
- **Min** moves the cursor to the minimum Y value.
- **Max** moves the cursor to the maximum Y value.

(Continued on next page)

Probe Note 9.1 (Continued)

The *search* command allows us to locate a particular point on the curve based on a number of conditions. To perform the search command: **Tools**, **Cursor**, **Search Commands** to open up the *Search Command* dialog box, enter a command, **OK**.

The search command is quite involved and will seldom be needed by the activities of this text. Therefore, a single example will suffice here.   (A complete description of the search command is found in the PSpice Circuit Analysis manual from MicroSim.)

sf/Begin/#2#(10ms,30ms,6,12)2:pe

This command (entered into the *Search Command* dialog box) means: search forward (*sf*), starting from first point in search range (*/Begin/*), require the two data points on each side of the target point to have lower Y-axis values (*#2#*), within the X-axis range of 10ms to 30ms and the Y-axis range of 6V to 12V (*10ms,30ms,6,12*), for the second occurrence (*2:*), of a peak (*pe*).

When the preceding command was performed on the rectifier curve of step 5 (after cursor 1 was placed at the lower left portion of the curve), the result is shown in Figure 9.5: Cursor 1 moved to the second peak within the search range, and there are at least two lower data points on either side.

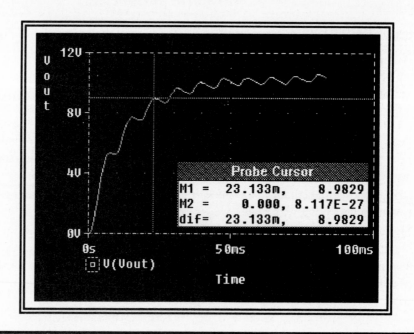

FIGURE 9.5

Using the search
command

Low-pass filter

7. Still we are not satisfied with the output. The ripple voltage is too large—yet we don't wish to increase the size of the expensive and bulky 100μF capacitor. The solution is to add a *low-pass filter*, which passes the DC voltage and shorts the AC ripple to ground. Make the necessary changes and create the circuit of Figure 9.6.

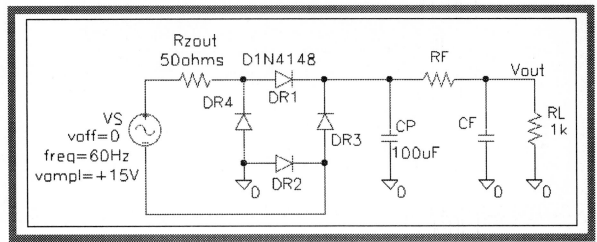

FIGURE 9.6

Power supply with low-pass filter added

8. Our next task is to choose values for RF and CF. Again, advanced mathematics would be required for a highly accurate answer. As before, estimated values will give satisfactory results. Using the following design guidelines, determine the required value of C:

- RF cannot be too large compared to the load (RL) because it would drop too much voltage. We arbitrarily choose 250Ω.

- CF must have a reactance (X_c) at the ripple frequency of 120Hz that is small when compared to RF. We arbitrarily choose a 10-to-1 ratio, giving X_c = 25Ω. Therefore:

    $1 / 2\pi fC$ = 25Ω, where π = 3.14 and f (ripple frequency) = 120Hz

    Giving, C = approximately _____

9. Assign to RF and CF the values determined in step 8 (250Ω and approximately 50µF?). Generate the new (filtered) Vout of Figure 9.7, add your curve to Figure 9.3, and tag it *"Filter added."*

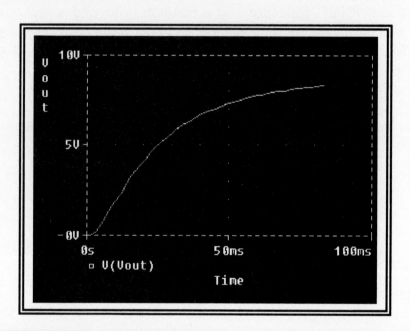

FIGURE 9.7

Filtered output waveform

10. Determine the ripple voltage at the far right of the curve as it approaches steady-state conditions. (Be sure to factor out as best you can the "background" slope of the curve.)

$V_{ripple} =$ _____

Has the ripple been reduced by some 90%?

    Yes   **No**

Voltage regulator

11. The ripple is now quite low, but the output voltage is nowhere near our 4.7V design goal. Furthermore, the output is not regulated because Vout varies considerably as the load (RL) changes.

    To solve both problems, add the zener regulator circuit developed in Chapter 8. As an added bonus, we will find that the ripple voltage is further reduced. Our final design is given in Figure 9.8.

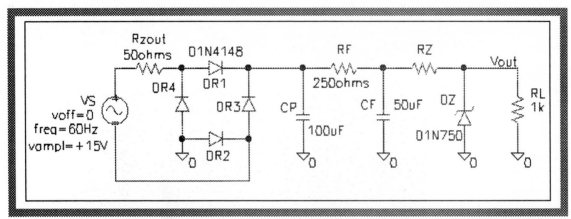

FIGURE 9.8

Final power supply
design with voltage
regulation

12. To determine RZ, we start with the ideal zener current (20mA) and work backward:

   • *The current through RZ*: From Chapter 8, the approximate value of current that gives a zener voltage of 4.7V is 20mA. The 1kΩ load adds some 5mA, for a total of 25mA through RZ.

   • *The voltage across RZ*: Starting with the original 15V, we subtract 1.75V for the diodes, 6.25V for the 25mA flowing through the 250Ω filter resistor, and 4.7V across the zener diode.   This leaves approximately 2.3V across RZ.   Therefore:

   RZ = 2.3V / 25mA = _____

13. Set the value of RZ according to step 12 (100Ω, after rounding off to the nearest "available" resistance value?).

14. Generate the final output curve of Figure 9.9 and add to Figure 9.3. Label this last curve "Regulator added." Also, determine the following by using steady-state values to the far right of the curve:

   Vout = _____

   Vripple = _____

   Is the output voltage near 4.7 and is the ripple very small?

   Yes   No

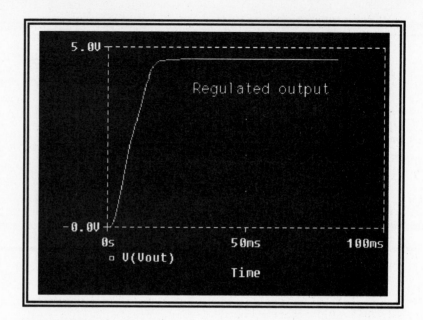

FIGURE 9.9

Regulated output

Regulation test

15. To test the voltage regulation characteristics of our design, generate the family of curves of Figure 9.10 by adding a nested sweep of the load value (RL).   (If necessary, review the *Process Summary* below.)

> **Process Summary for Regulation Test**
>
> • The main sweep variable is time, generated by a transient sweep from 0 to 83ms   The nested sweep variable is RL, generated by a parametric sweep from 200Ω to 1kΩ in increments of 200Ω. (See Chapter 5 for full details.)
>
> • The X-axis is time (primary sweep default), and the Y-axis is Vout [V(Vout)].

16. Based on your results:

   • At what approximate value of RL does the system appear to come out of regulation?

      Minimum RL for regulation = _____

   • How good is the regulation when RL is near 1kΩ? (Suggestion: Zoom in on the regulated output and determine ΔVout when RL goes from 800Ω to 1kΩ.)

      $\Delta V_{out} / \Delta RL = $ _____

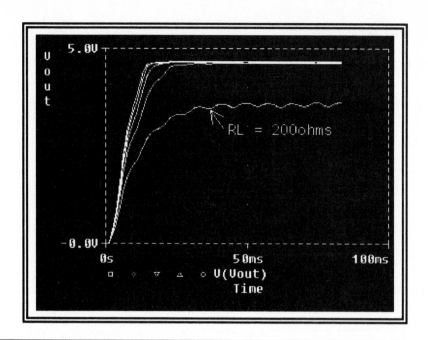

FIGURE 9.10

Regulated output
as a function of RL

Surge current

17. To the curves of Figure 9.10, add a graph of current through any
     of the rectifier diodes.   (Add a second Y-axis or second plot.)
     You will find that the highest current occurs during the initial
     (surge) cycle.

        Record this *maximum surge current* (ISurge) below and
     compare with the spec sheet value (*nonrepetitive peak surge
     current*) of 50mA.

             ISurge = _____

18. To lower the surge current, add a small *surge resistor* (RS) to
     your circuit (between the output of the bridge rectifier and CP).
     Determine its value by using the worst case equation below:

      VS (peak) / I (nonrepetitive peak) = 15V / 50mA = _____

        Measure the maximum surge current again.   Is it now below
     the spec sheet value?

             Yes   **No**

Advanced Activities

19. Add a front-end transformer (from library *analog.slb)* to your power supply, as shown in Figure 9.11. In addition to its output voltage, measure the primary and secondary voltages [vp and v(vs1,vs2)] and verify the turns ratio. (Because the power comes from the wall socket, Zout is very small.)

> Note: Should you encounter a "convergence" problem during calculation (time step goes below the minimum allowed value of E-15), force the system to take larger minimum steps by setting the *Step Ceiling* in the transient analysis to an appropriate value (say, 5μs).

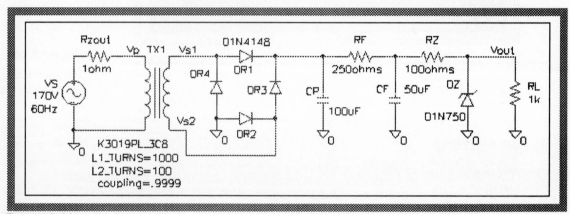

FIGURE 9.11

Adding a
transformer

20. For each of the following malfunctions, predict the approximate output voltage waveform. Make the change to the circuit of Figure 9.11 and compare to the PSpice-generated result.

   (a) Diode DR1 opens

   (b) CP opens

   (c) RZ shorts

EXERCISES

• Perform a heat analysis on various components (Rs and diodes) in the circuit. Compare to spec sheet and rated values. (Hint: Graph V*I or I^2R.)

QUESTIONS & PROBLEMS

1. Besides the power supply and load, what two components are required for a peak rectifier?

2. Fill in the blanks below with *resistor* or *capacitor*.

    With an RC low-pass filter, the DC component appears across the _____ and the AC component (ripple) appears across the _____.

3.   Referring to Figure 9.5, circle all the following processes that will decrease the ripple:

    (a) increase CP
    (b) decrease CP
    (c) increase CF
    (d) decrease CF

4. Besides regulating the voltage, why does the zener voltage regulator circuit also further reduce the ripple? (<u>Hint</u>: How does the zener RZ combination act as a voltage divider?)

5. Why is an LC filter more efficient than an RC filter?

6. Why is the surge current greatest during the first cycle?

CHAPTER 10

Clippers, Clampers, & Multipliers
Component Initialization

OBJECTIVES

- To design and analyze a variety of clippers, clampers, and multipliers.
- To initialize components in order to more quickly reach steady state.
- To generate *marching* waveforms.

DISCUSSION

Three of the most common applications of the diode are *clippers*, *clampers*, and *multipliers*. They are defined as follows:

- A *clipper* is a combination of diodes and resistors that limits the magnitude of a time-domain waveform.

- A *clamper* is a series combination of a diode and capacitor that adds a DC component to a time-domain waveform.

- A *multiplier* is a combination of diodes and capacitors that yields a DC voltage that is a multiple of the peak input voltage.

COMPONENT INITIALIZATION

Quite often we are interested in a circuit's steady-state response. However, based on the results of the power supply of Chapter 9, a great deal of computing time is often needed just to reach steady state. A case in point will be the multiplier circuit of this chapter. One solution is to *initialize* the capacitors to full charge before computation begins.

MARCHING WAVEFORMS

So far, all Probe-generated waveforms have been drawn at one time after all calculations were complete. In some cases, we can generate *marching waveforms*, which plot consecutive segments of the waveform while the calculations are under way. During time-consuming calculations, this gives us an opportunity to view results as they are available and to terminate simulation early.

Since our calculations are short, we will use the marching waveform feature to give a degree of "animation" to our waveforms.

SIMULATION PRACTICE

Clippers

1. Figure 10.1 shows a simple clipper and a *biased* clipper. In each case, predict the output time-domain waveforms for a +5V sine wave input. Sketch your predictions on the graphs of Figure 10.2.

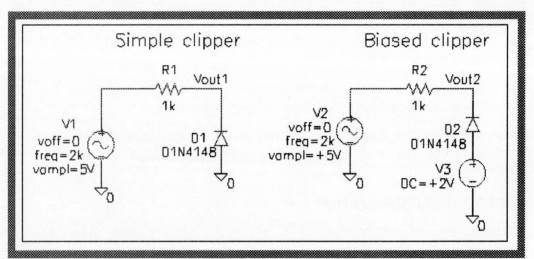

FIGURE 10.1

Clipper circuits

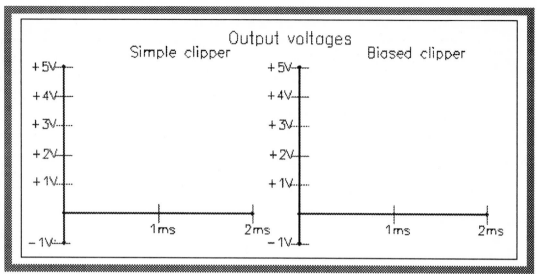

FIGURE 10.2

Clipper output
voltages

2. Using PSpice, draw the circuits and generate transient output
    waveforms.   (<u>Suggestion</u>: To speed the process and to make
    circuit comparisons easier, draw both circuits on the same
    schematic.)

3. Add the PSpice-generated waveforms to the graphs of Figure
    10.2. Clearly label all curves. Did your predicted curves match
    the experimental (PSpice) curves?

    Yes   **No**

Clampers

4. Figure 10.3 shows a simple clamper and a biased clamper. In
    each case, predict the output waveforms and sketch your
    predictions on the graphs of Figure 10.4.

5. Using PSpice, generate output waveforms and add them to the
    graphs of Figure 10.4. Did your predictions match the
    experimental curves?

    Yes   **No**

PSpice for Windows

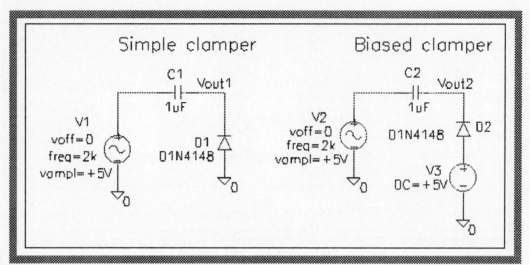

FIGURE 10.3

Clamper circuits

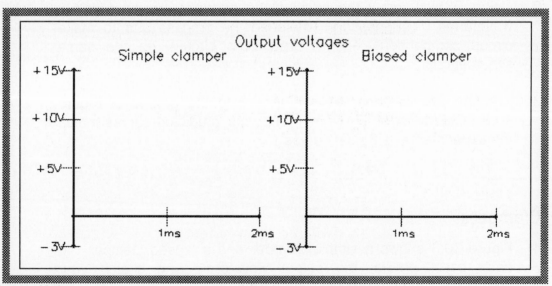

FIGURE 10.4

Clamper output
waveforms

PSpice for Windows

Multipliers

6. Figure 10.5 shows a common form of multiplier. In essence, it is a positive clamper followed by a peak rectifier. Predict the output voltage waveform and sketch on the graph of Figure 10.6. (<u>Hint</u>: It requires a large number of cycles to "pump up" the capacitors.)

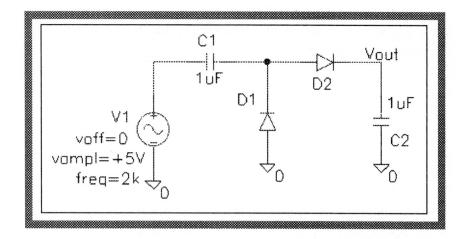

FIGURE 10.5

Multiplier circuit

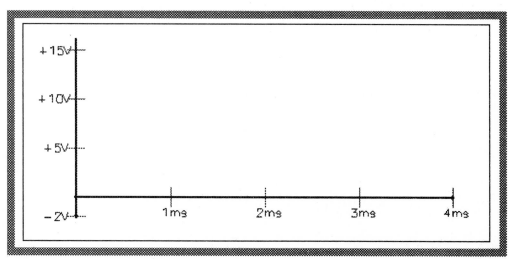

FIGURE 10.6

Multiplier waveform

PSpice for Windows

7. Using PSpice, generate the output waveform and add to the graph of Figure 10.6.

8. Chances are that step 7 revealed large differences between the predicted and actual (PSpice) waveforms. This is because the capacitors must be "pumped up" during the earlier cycles to approach their steady-state values.

    Using the *No-Print Delay* option (*PSpice Note 3.1*), generate the waveform of Figure 10.7, which shows the output from approximately 50 to 55 cycles (25ms to 27.5ms) as it approaches steady state. (<u>Hint</u>: Within the transient dialog box, set *Final Time* to 27.5ms and *No-Print Delay* to 25ms.)

- **What is the average value of Vout between 25ms and**

    27.5ms? _____

- **Is Vout still increasing?** _____

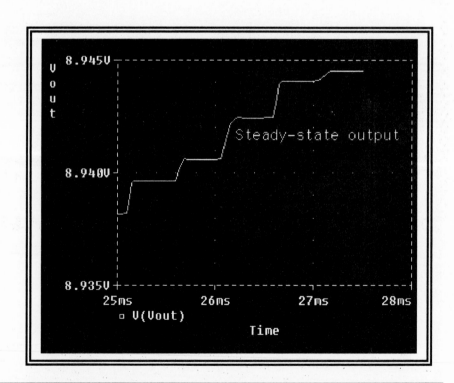

FIGURE 10.7

Steady-state output

Component Initialization

9. To more quickly approach steady-state conditions in the multiplier circuit of Figure 10.5, we initialize C1 and C2.

    Follow the directions of *Schematics Note 10.1* to pre-charge capacitor   C1 to 4.4V and C2 to 8.7V, as shown by Figure 10.8. (Use 4.4V and 8.7V because of the barrier potentials of D1 and D2.)

Schematics Note 10.1
How do I initialize a component value?

PSpice offers two methods for initializing components. Method one is the preferred technique.

* Method One: *Initial Condition Probes (IC1 and IC2)*

    IC1 is a probe that sets any circuit node to a selectable voltage.
    IC2 is a probe that sets any voltage *difference* to a selectable value.

    To set a probe: **Draw, Get New Part, Browse, special.slb, IC1 (or IC2), OK, Drag** to desired location, **CLICKL** to set, **CLICKR** to abort.   (**Edit, Flip** and **Edit, Rotate** as necessary.)

    To change the default initial condition of 0, **DCLICKL** on 0 and fill in the desired initial value, **OK**.

* Method Two: *Part Name*

    DCLICKL on component's *symbol* to bring up the Part Name dialog box, **IC=**, fill in the Value box, **Save Attr** (**Change Display**, etc., if desired), **OK**.   To enable the initialization: **Analysis, Setup, Transient, Use Init. Conditions, OK, Close**.

---

10. Change the transient display back to 0 to 4ms, display the new waveforms, and compare with the results of Figure 10.6. Does the circuit approach steady state sooner?

    Yes   **No**

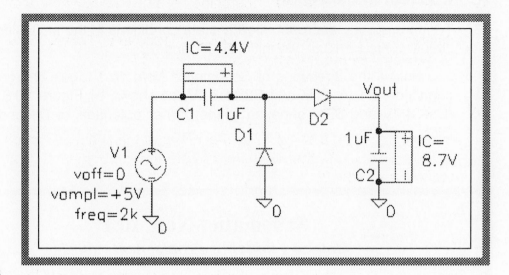

FIGURE 10.8

Component
initialization

Marching waveforms

11. Delete the initial condition devices (IC), and place voltage markers on the input and output in the multiplier circuit of Figure 10.8.

12. To enable the marching feature: **Analysis**, **Probe Setup**, **Monitor Waveforms [auto update]**, **Show All Markers** (so "marked" values will be displayed automatically), **OK**.

13. Run the simulation and note the waveforms generated.

14. Chances are, the waveforms did not "march" very smoothly. To improve the display: **Tools** (from Probe), **Options**, **Every %**, change 10 to 1 (for 1%), **Save**, **OK**. (The 1% entry means that a waveform segment will be added each 1% of the total calculation time.)

15. Run the simulation again and note the smoother waveform generation. (As the waveforms "march," note the percent display at the top of the Probe window.)

16. When done, disable the "marching" feature because it may interfere with other analyses: From Schematics, choose **Analysis**, **Probe Setup**, **Automatically Run Probe after Simulation**, **Restore Last Probe Session**, **OK**.

Advanced Activities

17. Using PSpice, do a *surge current* analysis of the multiplier circuit of Figure 10.5. Do the numbers indicate that a surge-protection resistor is needed?

EXERCISES

- Design a digital-based clipper (limiter) circuit that limits an input waveform to the range from 0 to +5V.

- By using two back-to-back (mirror image) multipliers (doublers) of Figure 10.5, design a multiplier that increases the peak input voltage by a factor of four.

QUESTIONS & PROBLEMS

1. What two components are necessary for clamping?

2. Show how to use a silicon diode (barrier = .7V) and a germanium diode (barrier = .3V) to create a clipper that limits an input waveform to the 0 to +1V range.

3. When voltage is "multiplied," what happens to the current? Why?

4.   Quite often, clamping is unwanted.   Referring to the circuit shown below, how does resistor R3 reduce the effects of clamping?

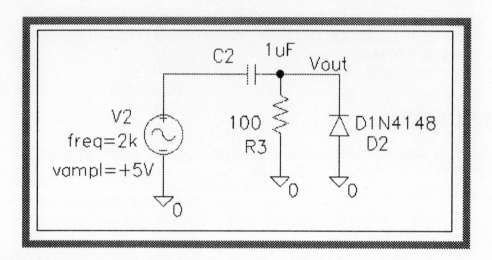

5.   Referring to Figure 10.5, why must capacitor C2 be "pumped up"?   (Why does it take an infinite number of cycles for C2 to reach full charge?)

CHAPTER 11

The Analog Switch
Crash Studies

OBJECTIVES

- To perform crash studies using a voltage-controlled switch.
- To use a diode to simulate the effects of safety devices.

DISCUSSION

This chapter's simulation study involves automotive safety. One of the most dangerous situations results from a quick stop (a crash). A crash is especially dangerous because of the great forces that can build up —even at moderate speeds. To model a sudden stop, we use the circuit of Figure 11.1.

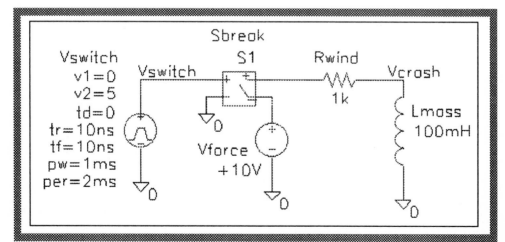

FIGURE 11.1

Analog computer circuit for crash studies

PSpice for Windows

Component S1 is a voltage-controlled switch. When the switch closes, Vforce is applied to Lmass and the car builds up velocity (current) until all the Vforce is used to overcome wind resistance (Rwind × I). When top speed is reached, maximum "kinetic" energy is contained in Lmass ($1/2LI^2$). When the switch opens (a crash occurs), the velocity is suddenly forced to zero and the back EMF ($V = L\Delta I/\Delta t$) simulates the very large forces.

To counter the large forces developed in a crash, we must simulate the use of safety devices, such as seat belts and air bags. To model such safety devices, we use a diode to separate the "speed-up" (accelerate) portion of the simulation from the "slow-down" (crash) portion.

SIMULATION PRACTICE

1. Draw the test circuit of Figure 11.1 and enter all parameters and attributes. [Switch S1 (Sbreak) is from the *Breakout.slb* library.]

2. Run a transient solution and generate the curves of Figure 11.2. [Note the minus sign in front of I(Lmass).]

   (a) Does the car reach steady state velocity prior to the crash?

         Yes   **No**

   (b) Which component *stores* energy prior to the crash?

          R   **L**

   (c) How large are the crash forces (back EMF voltage) and when do they occur?

3. Figure 11.2 has shown us what happens when an object is stopped suddenly, with no restraining devices. To simulate the effects of an air bag, add the resistor/diode circuit of Figure 11.3. The back EMF current now has a pathway through Rbag and the forces will be dissipated.

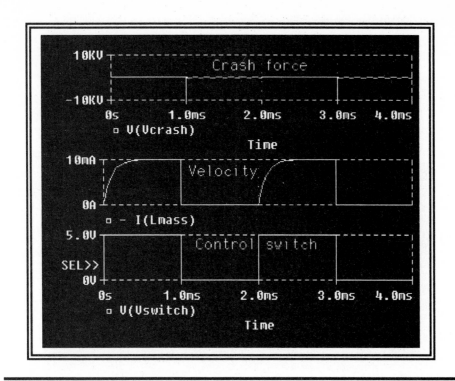

FIGURE 11.2

Crash study
results

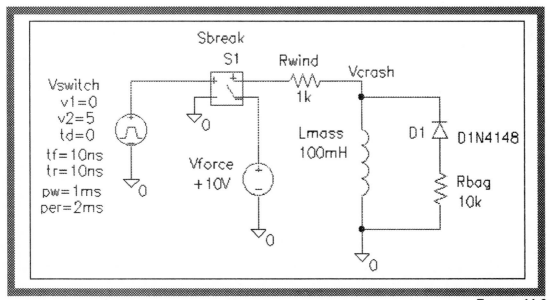

FIGURE 11.3

Simulating a
safety device

4. Test our new restraining system by generating the curves of Figure 11.4.   Are the crash forces greatly reduced?

   Yes   **No**

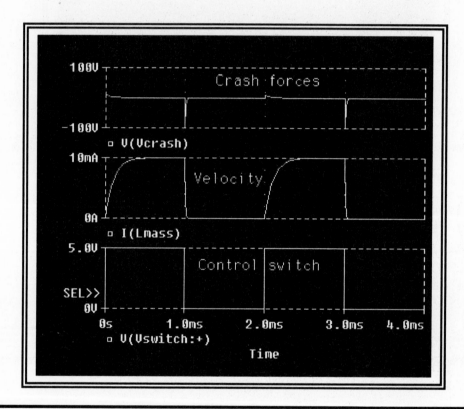

FIGURE 11.4

Restraining system
curves

5. To better interpret the forces that occur during the moment of crash, we would like to zoom in on just the *crash force* plot, leaving the *control switch* and *velocity* plots as they are.

   One method is to uncouple (unsync) the crash force plot and give it an X-axis time range that is independent of the other plots.

   Review *Probe Note 11.1* to uncouple the crash force plot, use any technique to zoom in on the crash force spike, and generate the waveform set of Figure 11.5.

PSpice for Windows

Probe Note 11.1
How do I uncouple individual plots of a multiple-plot graph?

When multiple plots are first generated, their X-axes are synchronized, and a zoom or plot action on one will affect them all.

To uncouple (unsync) a given plot: **CLICKL** on plot to select, **Plot**, **Unsync Plot**.   We are then free to use any Zoom or Plot technique on the uncoupled plot, independently of all the others.

As always, be sure to select the plot you wish to modify or examine by **CLICKL** on plot to place SEL>>.

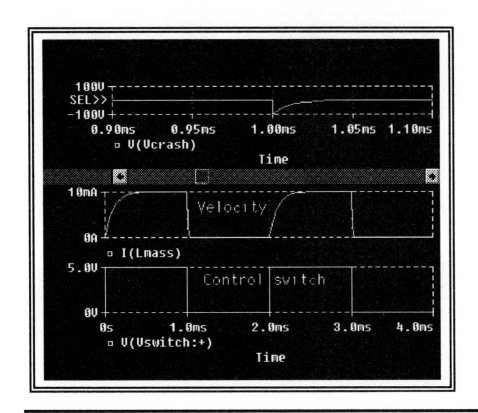

FIGURE 11.5

Unsynched crash force plot

Advanced Activities

6. To the graph of Figure 11.5, create a new window (**Window**, **New**) and add an expanded view of the crash details [V(Vcrash) and I(Lmass)], as shown in Figure 11.6   What is the relationship between velocity (current) and force (voltage across L)?

> Note: Figure 11.6 demonstrates all of the multiple-curve display methods available under *Probe*: multiple plots, multiple Y-axes, multiple windows, and unsynched plots. Be aware than SEL refers to a selected plot, and >> refers to a selected Y-axis. **CLICKL** on plot or Y-axis respectively to select.

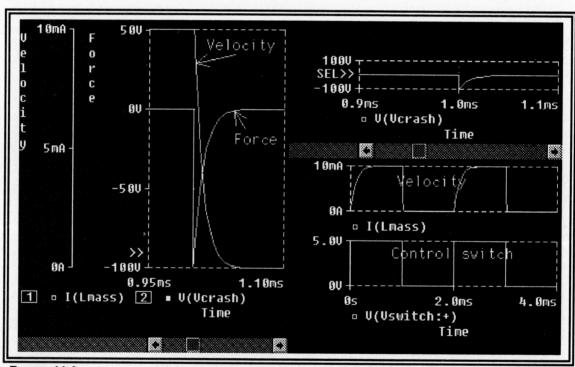

FIGURE 11.6

Multiple-plot
demonstration

7. Repeat the crash studies using the "open/close" switches of Figure 11.7. (Switches U1 and U2 are parts *Sw_tClose* and *Sw_tOpen* from library *eval.slb*.)

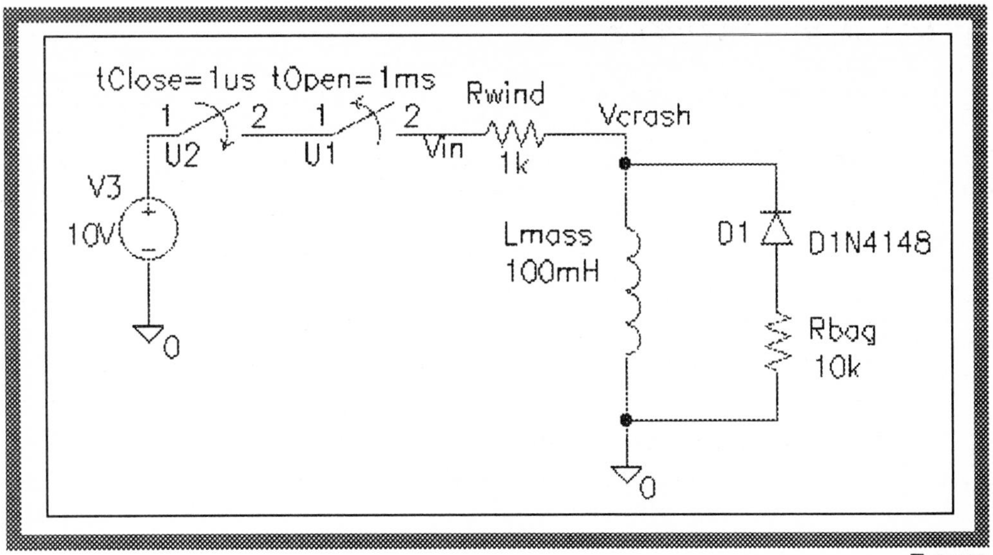

FIGURE 11.7

Crash study
using switches

8.   Plot graphs of *s(s(V(Vcrash)/100mH)))* and *s(I(Lmass))*.   Why are
they the same, and why are they an important factor in the design
of a restraint device? (Hint: Refer to the following equation, which
gives the distance traveled by an applied force—such as when an
object hits the restraint device.)

$$\text{Distance} = \iint \frac{Force}{Mass} \; dtdt = \int Velocity \; dt$$

EXERCISES

• Perform a comparison energy study of both crashes (with and
without Rbag). *Suggestion*: For both cases, determine the total
energy dissipated by Rbag, and compare to the "kinetic energy"
$(1/2LI^2)$.   Repeat using different values of Rbag. [Hint: Use the "s"
(integrate) operator to determine the area under the power curve—
which equals energy.]

QUESTIONS & PROBLEMS

1. Based on the results of this chapter, does switch S1 "bounce"? How could bouncing be simulated? (Hint: We have total control over the opening and closing of the switch.)

2. When Rbag is not present (Figure 11.1), where does the inductor (kinetic) energy go when a "crash" occurs?

3. To make the restraining system more effective (lower the crash forces still more), should Rbag (presently 10kΩ) be increased or decreased?   Why?

4. Looking to Figure 11.5, when the crash force is maximum what is true about the current?   (Hint: V = LdI/dt.)

5. Which of the following equations demonstrates Newton's second law of motion [F = ma, where a (acceleration) = dVelocity/dt].
   (a) V = LdI/dt
   (b) E = 1/2LI^2
   (c) Q = CV

6. What does our "crash study" tell you about the dangers of back EMF (the voltage caused by the presence of an inductor in a circuit that opens suddenly)?

PART III
The Bipolar Transistor

In Part III, we move to bipolar transistors, one of the fundamental building blocks of electronics. The emphasis is on amplifiers and buffers.

In Part III we take another step in turning over the PSpice setup process to the student. The "process summaries" will be used more sparingly, and it will be primarily the responsibility of the student to set up PSpice and generate the proper graphs. Just remember, to generate a Probe graph, you must answer the following two questions:

1. What sweep mode will I use?
2. What variables will I assign to the X- and Y-axes?

CHAPTER 12

Bipolar Transistor Characteristics
Collector Curves

OBJECTIVES

- To display collector and base curves for a *bipolar* transistor.
- To determine a transistor's *Beta*.
- To determine temperature effects.

DISCUSSION

A bipolar transistor is a solid-state device with two PN junctions. When *biased* as shown in Figure 12.1, it becomes a *voltage-controlled current source* with a current gain (*Beta*) in the range of 25 to 1000.

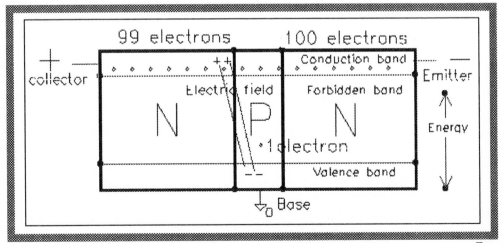

FIGURE 12.1

NPN transistor physical configuration

PSpice for Windows

VCIS

In simple terms, a bipolar transistor is useful in analog applications because it is a *voltage-controlled current source* (VCIS).   That is, the input "master" voltage (V_{BE}) controls an output "slave" current (I_C)— regardless of the "slave" voltage (V_{CB}).

Although it is not necessary to know *how* a transistor achieves its VCIS characteristics, observe the following from Figure 12.1:

- The "slave" (collector-base) circuit is a reverse-biased PN junction that acts as a charged capacitor with an electric field between its "plates." Therefore, no output current (I_C) flows as a result of output voltage (V_{CB}).   (The output current is independent of the output voltage.)

- The "master" (base-emitter) circuit is a forward-biased PN junction that "sprays" electrons between the plates of the capacitor according to the input voltage (V_{BE}).   In a typical case, 99% of these electrons are swept up by the electric field and become collector current (I_C).   Only 1% "fall" into the base and become base current (I_B).

When these two physical facts are put together, a transistor becomes a VCIS. (Give me an input voltage, and I will give you an output current—the output voltage doesn't matter.)

VCIS VERSUS ICIS

Because the "master" circuit of a bipolar transistor is a forward-biased PN junction, the driving source must supply current to the transistor. This is why a bipolar transistor is also called a *current-controlled current source* (ICIS).

TRANSISTOR BETA

To measure the effectiveness of any control device, we determine *output* divided by *input*.   Using current as our input/output variables, the transistor of Figure 12.1 has a current gain (*Beta*) of 99/1 (I_C/I_B). Because *Beta* (*B*) typically lies in the 25 to 1000 range, and because $I_E = I_B + I_C$, *B* is also approximately equal to I_E/I_B.

To investigate the VCIS/ICIS characteristics of a bipolar transistor, we will generate collector ("slave") curves. This is accomplished with the test circuits of Figure 12.2.

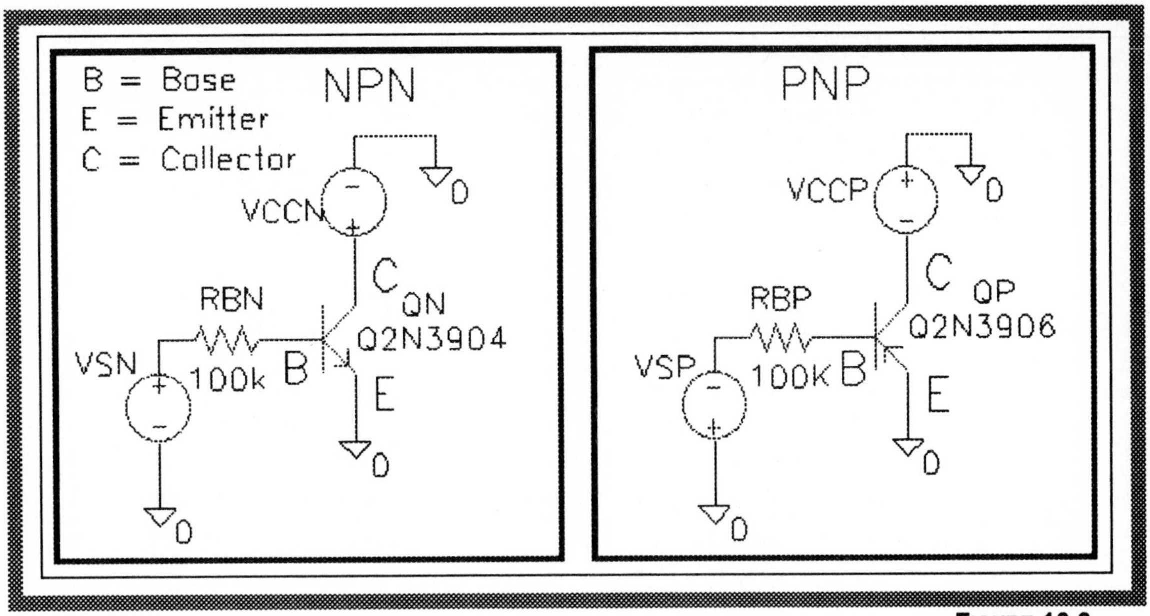

FIGURE 12.2

The bipolar transistor

NPN VERSUS PNP

The test circuits of Figure 12.2 show the schematic symbols for both the NPN and PNP versions. The PNP transistor performs the same as the NPN, except that all voltages and currents are reversed. (Note the change in direction of the arrow between the base and emitter.)

SIMULATION PRACTICE

Collector Curves (NPN)

1. Draw the NPN circuit of Figure 12.2 and set the attributes as shown. (Because both voltage sources will be swept, it is not necessary to assign them DC bias point values at this time.)

2. Using PSpice, generate the collector curves of Figure 12.3. (If necessary, refer to the *process summary* below.)

> ### Process Summary for Collector Curves
>
> - The <u>Main Sweep</u> variable is collector voltage (VCCN), generated with a linear DC Sweep from 0 to +10V in increments of .1V. The <u>Nested Sweep</u> variable is VSN from 0V to +10V in linear increments of 2V.
>
> - The <u>X-axis</u> variable is V_VCCN (Main Sweep default), and the <u>Y-axis</u> variable is IC(QN) (collector current).

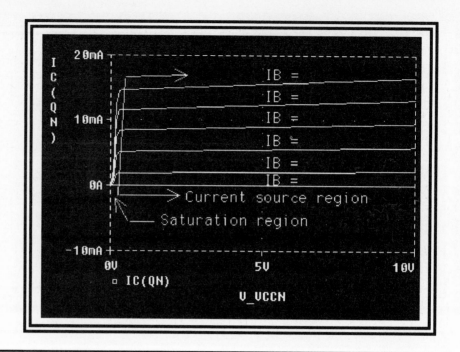

FIGURE 12.3

Collector curves

3. Add another Y-axis to your graph and generate a set of base current [IB(QN)] curves. Use the data to fill in the corresponding "IB =" for each of the six curves of Figure 12.3 (use the margin and arrows, if necessary).

4. State in your own words how the graph of Figure 12.3 proves that a bipolar transistor (in the current source region) is an ICIS. (<u>Hint</u>: Does a smaller I_B control a larger I_C, and why is it important that the curves are flat?)

5. Determine the collector impedance (Z_C) anywhere in the current-source region (flat portion of the curves) for the "top-most" and "bottom-most" curves. (<u>Hint</u>: Use the 1/slope method, as demonstrated for the top-most curve in Figure 12.4, or use the "d" operator.)

    Z_C **(top-most) = _____**

    Z_C **(bottom-most) = _____**

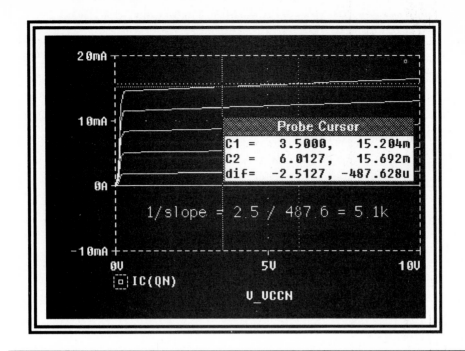

FIGURE 12.4

Determining collector impedance

6. Determine a *typical* value for *Beta* (I_C/I_B) in the current-source region for the top-most and bottom-most curves:

   Beta (top-most) = _____

   Beta (bottom-most) = _____

7. Examine the model parameters of the 3904 (select, **Edit**, **Model**, **Edit Instance Model**) and note the value of the *ideal maximum forward Beta* (Bf).

   Beta (Bf) = _____

   Comparing steps 6 and 7, is it fair to say that a *typical* value of *Beta* is much less than the *ideal maximum* value?

   Yes **No**

8. The region from V_C = 0 to .4V is known as the *saturation* region. This is because the low collector voltage causes electrons to "fall" into (and "saturate") the base, rather than being swept to the collector. In the saturation region, the transistor is no longer a current source. Determine a typical (*approximate)* value of *Beta* when V_C = .1V. (<u>Hint</u>: Expand the lower range of the curves.)

   Typical *Beta* (saturation region) = _____

Base ("Master") Curves

9. The collector curves of Figure 12.3 do not *directly* show the VCIS transconductance relationship. (How does the input *voltage* affect the output *current*?) To see this relationship, generate the "master" curve of Figure 12.5. (If necessary, refer to the *process summary* below.)

Process Summary for "Master" Bipolar Curve

- The <u>Main Sweep</u> variable is VSN, from 0 to +10V in increments of .1V. (The Nested Sweep is disabled and VCCN is set to +10V.)

- The <u>X-axis</u> variable is V_{BE} [V(QN:b)], and the <u>Y-axis</u> variable is I_C [IC(QN)].

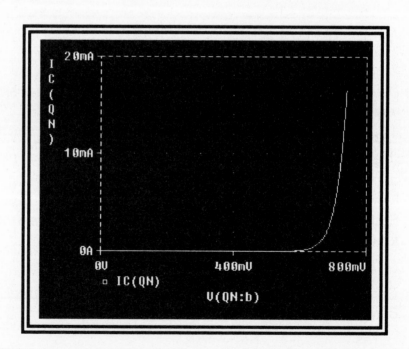

FIGURE 12.5

Collector current
versus base voltage

10. As expected, Figure 12.5 shows the characteristics of a forward-biased PN junction. Is it similar to the diode curve of Figure 8.5?

   Yes   **No**

Temperature Effects

11. Of all the temperature-related variables of a bipolar transistor, *Beta* (*B*) is the most sensitive. Recreate the graph of Figure 12.6, which shows how *B* changes with temperature. Note that the cursor is set at two arbitrary coordinates, showing a *B* of 122.9 at -40.1° and 189.1 at +40.1°. (If necessary, refer to the *process summary* below.)

Process Summary for Temperature Effects of *B*

- The <u>Main Sweep</u> variable is QN temperature, using a DC Main Sweep from -50 to +50 in increments of 1°. (The Nested Sweep is disabled; VCCN and VSN are both set to +10V.)

- The <u>X-axis</u> variable is temperature (Main Sweep default), and the <u>Y-axis</u> variable is I_C/I_B (*Beta*).

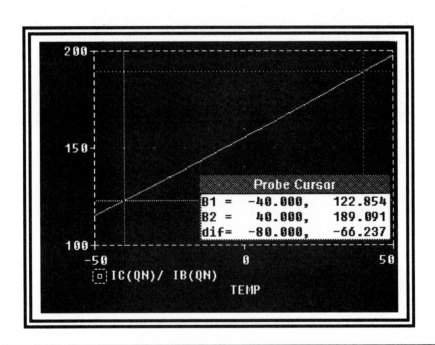

FIGURE 12.6

Beta as a function of temperature

12. Based on the results of step 11, *Beta* is quite dependent on temperature. As a quantitative measure of this dependency, determine the following:

   Δ**B / ΔTemp = slope = _____**

Advanced Activities

13. Generate a set of collector curves for the PNP circuit of Figure 12.2. Based on your results, what is the difference in operation between an NPN and PNP transistor?

14. Change the *ideal maximum Beta* of the transistor (model parameter *Bf*) and generate new collector curves. Compare to Figure 12.3.

EXERCISES

- Design a current source of 10mA using each of the following:

    (a) a 3904 NPN transistor
    (b) a 3906 PNP transistor

QUESTIONS & PROBLEMS

1. What is a *current source*?

2. If for every 100 electrons that are pulled into the emitter, 98 go on to the collector, what is the *Beta?*

3. With the emitter grounded, what is the approximate minimum collector voltage that will result in "normal" (nonsaturated) operation?   (Hint: See Figure 12.3.)

4. For the circuits of Figure 12.2, place "base-emitter" or "collector" in the correct spaces below.

    VCIS means that the input _____ voltage controls the output _____ current, regardless of _____ voltage.

5. A transistor's *alpha* is equal to I_C / I_E.   If *Beta* for a given transistor is 200, what is *alpha*?

6. Based on Figure 12.3, what is a *typical* (average) value for Z_C (collector impedance in the current-source region)?

7. When the temperature rises *Beta*

    (a) goes up
    (b) goes down

8. When using the transistor schematic symbol, the arrow points in the direction of emitter

    (a) electron flow
    (b) conventional flow

CHAPTER 13

Bipolar Biasing
Stability & DC Sensitivity

OBJECTIVES

- To design and analyze several popular bipolar transistor biasing circuits.
- To compare the temperature stability characteristics of biasing circuits.
- To perform a DC sensitivity analysis on bias circuits.

Discussion

In analog applications, a bipolar transistor is primarily used as an amplifier or buffer. Because analog applications usually involve an AC signal, we must *bias* the transistor. (As we will see in Chapter 21, a bipolar transistor used in a digital application is a *switch* and does not require biasing.)

To bias a bipolar transistor is to use a DC voltage to place its *quiescent* (Q) *point* at an appropriate place in the "master" curve. When properly biased, the superimposed AC signal will have room to operate on both its positive and negative cycles. A typical Q point for the 3904 NPN bipolar transistor is shown in Figure 13.1.

Because of a transistor's sensitivity to voltage (beyond the .7V "knee"), Figure 13.1 shows that we usually establish the Q point by designing for the desired *current* (such as 10mA).

There are many trade-offs involved in the design of a bias circuit. Several of the most important are *cost*, *temperature stability*, *sensitivity* to tolerances, and *number of power supplies required*.

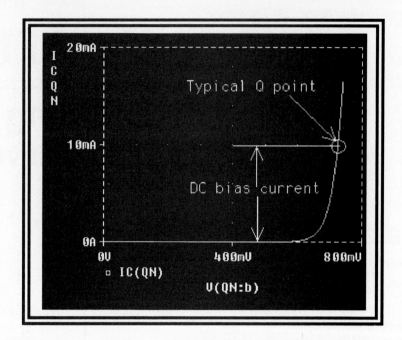

FIGURE 13.1

NPN transistor
biasing and Q points

DC SENSITIVITY

DC sensitivity analysis is useful during the design of a bias circuit. The question is, how sensitive is the Q point to a change in the value of various components?   For example, if we find that the Q point is very sensitive to a particular resistor, we might want to reduce that resistor's tolerance from 10% to 5%.

To perform a DC sensitivity analysis, we first select one or more output variables. By performing a linear analysis of all devices about the bias point, the sensitivities of each of the output variables to all the device values and model parameters will be calculated and sent to the *output file*.   (No Probe graph is generated.)

We may select any output variable for a sensitivity analysis; however, if we choose a current, it must be through a voltage source.

SIMULATION PRACTICE

Base Biasing

1. The simplest type of biasing for a bipolar transistor is the *base-biased* circuit of Figure 13.2.   The following equations govern the DC values:

    *   *Beta* = I_C / I_B = (approximately) I_E / I_B
    * V_{CC} = .7V + $I_B R_B$
    * V_{CC} = V_C + $I_C R_C$

    Assuming a *Beta* of 175, solve for the Q point by hand and fill in the following values. (C = "collector" and CEQ = "difference between collector and emitter".)

    I_C = _____ V_{CEQ} = _____

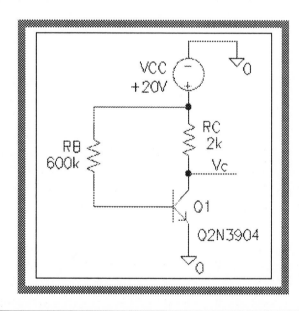

FIGURE 13.2

Base-biased circuit

2. Using PSpice, run a DC bias point solution (**iprobe** and **viewpoint**, or examine the output file) and determine the Q point values:

    I_C = _____ V_{CEQ} = _____

3. Compare the theoretical values of step 1 with the experimental values of step 2.   Are they approximately the same?

   Yes   No

Temperature Stability

4. Using just a single base resistor, the *base-biased* circuit is inexpensive—but is it stable? A good measure of stability is, what happens to the Q point current (I_{CQ}) as the temperature changes? To determine Q point stability, generate the graph of Figure 13.3. (If necessary, refer to the *process summary* below.)

Process Summary for Base-Biased Stability

- The <u>Main Sweep</u> variable is temperature, generated by a DC Sweep of Q1 from -50C to +50C in increments of 1C. (The Nested Sweep variable is disabled.)

- The <u>X-axis</u> variable is temperature (Main Sweep default), and the <u>Y-axis</u> variable is collector current [IC(Q1)].

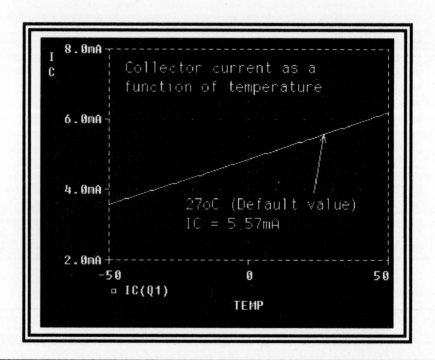

FIGURE 13.3

I_{CQ} as a function of temperature

5. Based on the results of step 4, it is clear that a base-biased circuit is not stable. As a quantitative measure of this instability, determine the slope of the curve.

   $\Delta I_{CQ} / \Delta Temp =$ _____

Voltage-Divider Bias

6. The most popular biasing circuit is shown in Figure 13.4. It uses *voltage-divider* biasing to further improve stability. Following the same steps as with base biasing, determine each of the following:

   - Using calculations: $I_{CQ} =$ _____ $V_{CEQ} =$ _____
   - Using PSpice: $I_{CQ} =$ _____ $V_{CEQ} =$ _____

     $\Delta I_{CQ} / \Delta Temp =$ _____

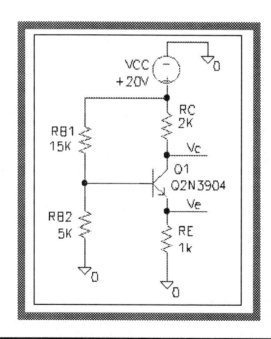

FIGURE 13.4

Voltage-divider bias

7. Based on the results of this chapter so far, why is voltage-divider biasing superior to base-biasing? (<u>Hint</u>: Is it more stable to changes in temperature?)

Collector-Feedback and Emitter Biasing

8. Two additional biasing circuits are the *collector-feedback* and *emitter* biasing circuits of Figure 13.5(a) and (b).   Use PSpice to analyze both circuits for temperature stability and enter your results below:

  - **Collector-feedback:** $\Delta I_{CQ} / \Delta$**Temp** = _____
  - **Emitter biasing:** $\Delta I_{CQ} / \Delta$**Temp** = _____

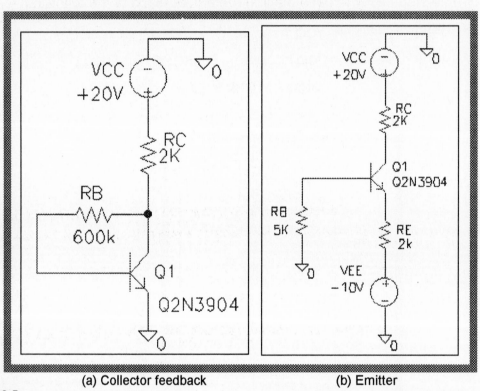

(a) Collector feedback   (b) Emitter

FIGURE 13.5

Additional bipolar
bias circuits
(a) Collector feedback
(b) Emitter

9. Based on the results of this chapter, rank the four types of bias circuits according to temperature stability. (Place 1, 2, 3, or 4 beside each.)

_____   **Base bias**   _____   **Collector-feedback bias**

_____   **Voltage-divider bias**   _____   **Emitter bias**

DC Component Sensitivity

10. Bring back the voltage-divider bias circuit of Figure 13.4.

11. Enable the *Sensitivity Analysis* mode, and open the dialog box of Figure 13.6 (**Analysis**, **Setup**, **CLICKL** on sensitivity enabled box, **Sensitivity**). Enter the desired variable we wish to examine for sensitivity [such as V(Vc,Ve), the Q point voltage *difference* between collector and emitter], **OK**, **OK**. (No other analysis need be checked as no Probe graph need be generated.)

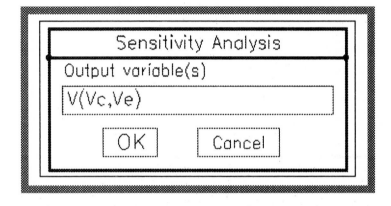

FIGURE 13.6

Sensitivity Analysis dialog box

12. Analyze the circuit (**Analysis**, **Simulate**), and examine the *DC Sensitivity Analysis* portion of the output file (Table 13.1). The top of the table indicates that all values refer to the Q point voltage [V(Vc,Ve)], the voltage difference between collector and emitter.

13. Next, we observe that the data (Table 13.1) is divided into two sections: resistor and voltage-source components at the top, and transistor parameters at the bottom.

    (a) How many R and V components affect V(Vc,Ve)? _____

    (b) How many transistor parameters affect V(Vc,Ve)? _____

DC SENSITIVITY ANALYSIS TEMPERATURE = 27.000 DEG C			
DC SENSITIVITIES OF OUTPUT V(Vc,Ve)			
ELEMENT NAME	ELEMENT VALUE	ELEMENT SENSITIVITY (VOLTS/UNIT)	NORMALIZED SENSITIVITY (VOLTS/PERCENT)
R_RC	2.000E+03	-4.160E-03	- 8.320E-02
R_RB1	1.500E+04	7.287E-04	1.093E-01
R_RB2	5.000E+03	-2.132E-03	-1.066E-01
R_RE	1.000E+03	7.994E-03	7.994E-02
V_VCC	2.000E+01	2.712E-01	5.424E-02
Q_Q1			
RB	1.000E+01	7.246E-05	7.246E-06
RC	1.000E+00	1.936E-05	1.936E-07
RE	0.000E+00	0.00E+00	0.000E+00
BF	4.164E+02	-2.989E-04	-1.244E-03
ISE	6.734E-15	2.904E+13	1.956E-03
BR	7.371E-01	1.597E-10	1.177E-12
ISC	0.000E+00	0.000E+00	0.000E+00
IS	6.734E-15	-3.421E+13	-2.304E-03
NE	1.259E+00	-3.346E+00	-4.213E-02
NC	2.000E+00	0.000E+00	0.000E+00
IKF	6.678E-02	-2.883E-01	-1.925E-04
IKR	0.000E+00	0.000E+00	0.000E+00
VAF	7.403E+01	4.211E-04	3.188E-04
VAR	0.000E+00	0.000E+00	0.000E+00

TABLE 13.1

DC sensitivity portion
of output file

14. To interpret the table, look at the first line.   It tells us that for every 1Ω change in the $2k\Omega$ value of R_C, V(Vc,Ve) would change by $-.00416V$.   Or, for every 1% change in R_C (20Ω), V(Vc,Ve) would change by $-.0832V$.

    As an example, if R_C changes from $2k\Omega$ to $2.1k\Omega$, enter the expected change in V(Vc,Ve) below: (Reminder: V(Vc,Ve) = V_{CEQ}.)

    ΔV_{CEQ} **(from table) =** _____

15. Using the method of your choice (viewpoint or output file), measure the DC collector/emitter voltage when R_RC = $2k\Omega$ and again when R_RC = $2.1k\Omega$. Enter the results below: Do they agree with those of step 14?

    ΔV_{CEQ} **(from PSpice) =** _____

16. By examining the data of Table 13.1, the output Q point voltage [V(Vc,Ve)] is most sensitive to a *percent* change in which resistor? Circle your answer:

    R_C R_{B1} R_{B2} R_E

17. From Table 13.1, if the transistor's *maximum Beta* parameter (Bf) changes by 10%, what change occurs in the output Q point voltage?

    $\Delta V_{CEQ} =$ _____

Advanced Activities

18. By changing the value of R_B, redesign the collector feedback circuit of Figure 13.5(a) to give *midpoint bias* operation (V_C = 10V).

19. Redesign any of the bias circuits of this experiment using a PNP (3906) transistor. (<u>Hint</u>: Following convention, draw the circuit upside down.)

EXERCISES

• Design a bias circuit giving an I_C = 10mA and having maximum temperature stability. (<u>Hint</u>: What configuration is best?)

QUESTIONS & PROBLEMS

1. What is the purpose of biasing a transistor? (<u>Hint</u>: What would be the result if the transistor were not biased?)

2. Based on the bipolar examples of this experiment, which of the following results in the most stable circuit   (R_B refers to the base resistors and R_E refers to the emitter resistor)?:

   (a) R_B small and R_E large

   (b) R_B large and R_E small

3. When biasing bipolar transistors, why is it easier to design for bias *current* (and let the bias voltage "tag along")?

4. What is the major disadvantage of emitter biasing [Figure 13.5(b)]?

5. Describe the feedback process in the collector-feedback circuit of Figure 13.5(a).   (Why is collector-feedback biasing more stable than base-biasing, although both use the same number of bias resistors?)

6. To improve the design of the voltage-divider bias circuit of Figure 13.4, you might lower the tolerance of two of the four resistors (for example, from 10% to 5%).   Based on the results of the sensitivity analysis (Table 13.1), which two resistors would you select?   Why?

CHAPTER 14

Bipolar Amplifier
Small-Signal

OBJECTIVES

- To analyze a common-emitter small-signal amplifier.
- To demonstrate the trade-off between linearity and gain.
- To examine a transistor's *model*.

DISCUSSION

The small-signal amplifier circuit of Figure 14.1 uses *voltage-divider* biasing for stability. Because the emitter is AC grounded, it is in the *common-emitter* configuration.

   To better understand the action of the circuit, we apply the *superposition theorem* to divide its operation into two parts: a DC *equivalent circuit* [Figure 14.2(a)], and an AC *equivalent circuit* [Figure 14.2(b)]. The total voltage or current at any circuit point is the algebraic sum of the DC and AC values. (DC variables are usually labeled with capital letters, and AC variables with lowercase letters.)

- **DC equivalent circuit**   The purpose of the DC equivalent circuit is to *bias* the transistor to its Q point. To extract the DC equivalent circuit of Figure 14.2(a), we open all capacitors, remove all AC sources, show the collector as a current source, and the base-emitter (BE) junction as a diode. We calculate the DC values as follows ("||" means "in parallel with"):

  - **DC *Beta* = 175 (assumed value)**

  - $I_{EQ} = I_{CQ} = (V_{th} - .7V) / (R_E + (R_{B1}||R_{B2})/B)$
    $(10V - .7V) / (2k\Omega + (10k\Omega||20k\Omega)/175) = 4.65mA$

  - $V_{CEQ} = V_{CC} - I_{CQ}(R_C + R_E) = 30V - 4.65mA(2k\Omega + 2k\Omega) = 11.4V$

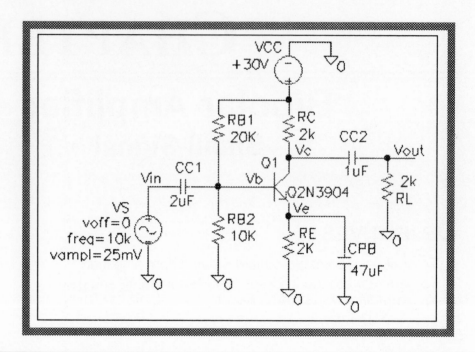

FIGURE 14.1

Common-emitter
small-signal amplifier

- **AC equivalent circuit** The purpose of the AC equivalent circuit is
  to amplify the signal. To extract the AC equivalent circuit of Figure
  14.2(b), we short all capacitors and DC power supplies, turn the
  collector into a current source, and show the BE junction as a bulk
  resistor. The *approximate* AC values are as follows:

  - **AC *Beta* = 175 (assumed to be the same as DC *Beta*)**

  - **Bulk resistance (known as re') $\approx$ 25mV / I_{EQ} = 25mV / 4.6mA = 5.4Ω**

  - **A (voltage gain) = $R_C \| R_L$ / re' = (2kΩ $\|$ 2kΩ) / 5.4Ω = 1kΩ / 5.4Ω =
    185**

  - **Input impedance (Z_{in}) = $R_{B1} \| R_{B2} \| (B \times$ re') = 10k$\Omega$$\|$20k$\Omega$$\|$(175 $\times$
    5.4Ω) = 900Ω.**

  - **Output impedance (Z_{out}) = $r_c \| Z_{current\ source}$* = 2k$\Omega$$\|$10k$\Omega$ =
    1.67kΩ**
    * $Z_{current\ source}$ is from Chapter 12, and assumed to be 10kΩ.

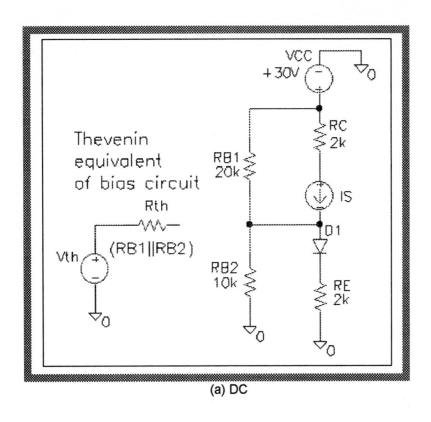

(a) DC

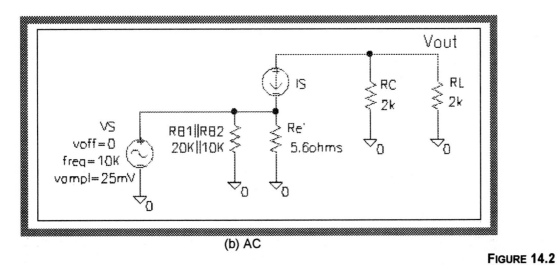

(b) AC

FIGURE 14.2

DC and AC
equivalent circuits
(a) DC
(b) AC

THE COUPLING AND BYPASS CAPACITORS

The coupling capacitors (CC1 and CC2) couple the AC in and out of the circuit without disturbing the DC biasing. The bypass capacitor (CBP) increases the voltage gain by shorting the AC signal to ground (bypassing RE). To perform their duties properly, the values of CC1, CC2, and CBP must be large enough to act as near shorts—but not too large, as they can be expensive and bulky.

As an example, let's see how the value of CC1 was determined. The first step is to Thevenize the amplifier and generate the equivalent circuit of Figure 14.3. We also assume (arbitrarily) that the lowest frequency of interest is 1kHz.

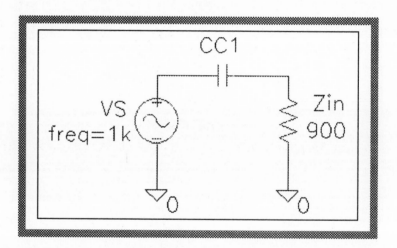

FIGURE 14.3

Equivalent circuit
to isolate CC1

Because the capacitive reactance (X_{CC1}) can never be zero (unless f or C is infinite), we must choose a reasonable value for X_{CC1}. With a Z_{in} of 900Ω, we arbitrarily choose a value of 90 (1/10 of Z_{in}). The equation then becomes:

$$X_{CC1} = 1 / 2\pi fC = 1 / (2 \times 3.14 \times 1k \times CC1) = 90\Omega$$

solving for CC1 yields approximately 2μF

Values for the other two capacitors were determined in the same manner.

EXPERIMENTALLY MEASURING Z_{IN} AND Z_{OUT}

Referring to the Thevenized version of the amplifier (Figure 14.4), we experimentally determine Z_{in} by dividing V_{in} by I_{in}. We determine Z_{out} by measuring V_{out} with and without a load. Without a load (RL = infinity), $V_{out} = V_{th}$; with a load, V_{out} is less than V_{th}. We then use algebra to determine Z_{out}.

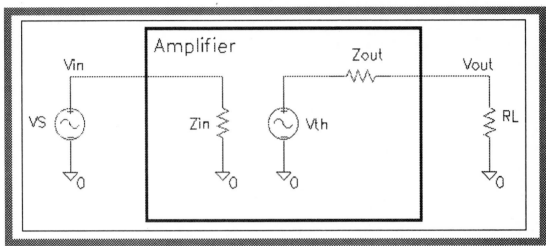

FIGURE 14.4

Thevenized amplifier, showing Z_{in} and Z_{out}

TIME DOMAIN VERSUS FREQUENCY DOMAIN

In this chapter, we use time-domain (transient) analysis to determine all amplifier characteristics—the type of analysis that might be done "hands-on" with an oscilloscope. In the next chapter, we will switch to frequency-mode analysis—the type of analysis that might be done "hands-on" with a spectrum analyzer. We will find that there are many advantages to including the frequency mode in any circuit analysis.

SIMULATION PRACTICE

1. Draw the circuit of Figure 14.1 and set the attributes as shown.

DC Analysis

2. Using PSpice, perform a DC bias-point analysis on the amplifier. (Set *viewpoints* and *iprobes*, or examine the *small signal bias solution* in the *output file*). Record each of the values listed below:

    $I_{EQ} = I_{CQ}$ **(approximately)** = _____ V_{CEQ} = _____

3. Compare your experimental DC results from step 2 with the theoretical predictions made in the discussion. Are they *approximately* the same (within 10%)?

    Yes   No

AC Analysis

4. Using PSpice, perform a transient analysis and determine the following: (<u>Note</u>: *Use peak-to-peak* values for all voltages and currents to average out the effects of distortion.)

    $A (v_{out} / v_{in})$ = _____
    AC *Beta* = i_c / i_b = _____
    $Z_{in} = v_{in} / i_{in}$ = _____
    Z_{out} = _____ **(See discussion on Z_{in} and Z_{out}.)**

5. Compare your experimental AC results from step 4 with the theoretical predictions made in the discussion. (Are they generally the same? Comment on any significant differences.)

6. Using the technique of your choice, determine if the coupling and bypass capacitors are "doing their job" (effectively shorting the AC signal). Summarize your findings. (<u>Hint</u>: Measure the *AC* voltage drop *across* the capacitors.)

Linearity

7. Look at the amplifier's output signal. Is it distorted (*nonlinear*)? Use the following equation to give a quantitative measure of the distortion. (See Figure 14.5 for an example.)

$$\% \text{ distortion} = \frac{\textbf{Vpeak(difference)}}{\textbf{Vpeak(average)}} \times 100 = \underline{\hspace{2cm}}$$

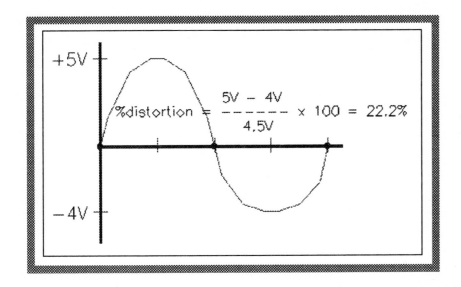

FIGURE 14.5

Example of percent distortion calculation

8. To reduce distortion (at the expense of gain), we add the 100Ω *swamping* resistor (R_S), as shown in Figure 14.6. (<u>Note</u>: We increase V_S to 250mV to compensate for the reduced gain.)

9. For the circuit of Figure 14.6, determine *theoretically* (by hand calculation) the "big three": A, Z_{in}, and Z_{out}.

    (a) $A = (R_C \parallel R_L) / (R_S + re') = \underline{\hspace{3cm}}$

    (b) $Z_{in} = R_{B1} \parallel R_{B2} \parallel B(R_S + re') = \underline{\hspace{3cm}}$

    (c) $Z_{out} = R_C \parallel Z_{collector} = \underline{\hspace{3cm}}$

10. Using PSpice, *experimentally* determine the "big three."

     (a) A = _____

     (b) Z_{in} = _____

     (c) Z_{out} = _____

11. Are the theoretical and experimental values generally the same? (Compare steps 9 and 10.)

     Yes   **No**

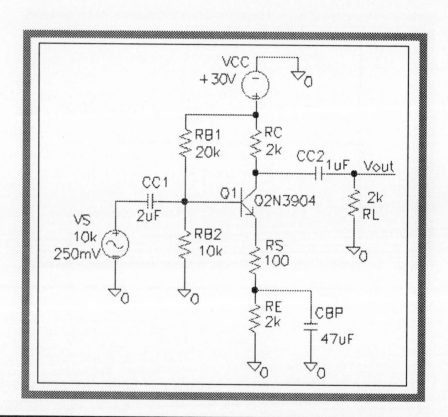

FIGURE 14.6

Swamped amplifier

12. Generate an output waveform, and determine the "swamped" degree of distortion.

$$\% \text{ distortion} = \frac{\textbf{Vpeak(difference)}}{\textbf{Vpeak(average)}} \times 100 = \text{_____}$$

13. Swamping produced what percent change in each of the following:

    (a) **Distortion reduced (%) =** _____

    (b) **Gain reduced (%) =** _____

The Initial Transient Solution

14. Plot Vc (the collector voltage) and examine it's amplitude at TIME = 0. Examine Vc in the INITIAL TRANSIENT SOLUTION in the output file.   Are they the same?

    Yes   No

Advanced Activities

15. Create the *common-base* (base is AC grounded) circuit of Figure 14.7 by simply moving V$_S$ from the base to the emitter. Measure A and Z$_{in}$. Why did A remain the same, but Z$_{in}$ go way down? (The common-base configuration is often used in high-frequency applications because of its reduced input capacitance.)

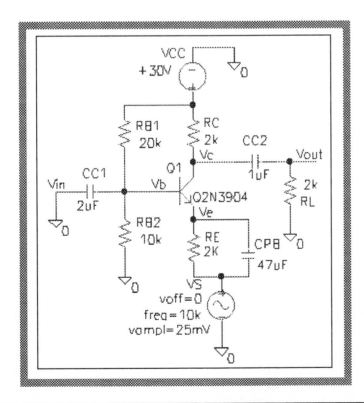

FIGURE 14.7

Common-base configuration

PSpice for Windows

16. Redesign the amplifier of Figure 14.6 using a PNP transistor. Are the major characteristics (A, Z_{in}, Z_{out}, linearity) the same as the NPN transistor circuit?

17. For either of the amplifiers of this experiment (swamped or unswamped), determine re' experimentally from the gain and the gain equation. Compare your result to the theoretical value (re' = 25mV/I_{EQ}).

EXERCISES

- Design a highly stable, highly linear *two-stage* common-emitter amplifier with an overall gain of 100.

QUESTIONS & PROBLEMS

1. Referring to Figure 14.6, what is the purpose or function of each of the following components?:

   (a) Capacitors CC1 and CC2

   (b) Capacitor CBP

   (c) Resistor RE

   (d) Resistor RS

2. To give an undistorted (linear) output signal, why is it important for the transistor to be a *current source*?

3. Why is the circuit of Figure 14.1 stable but nonlinear?

4. Regarding the amplifier circuit of Figure 14.1, what is the phase relationship between input and output?

5. Referring to Figure 14.1, if the lowest frequency of interest were decreased from 1kHz to 100Hz, should we increase or decrease the values of the coupling and bypass capacitors?

6. What is the *power gain* of the circuit of Figure 14.6?   (<u>Hint</u>: Power gain = A x *B*.)

7. What would happen to the voltage gain of the swamped circuit of Figure 14.6 if the bypass capacitor (CBP) opened?

8. What is the only difference between the *common-emitter* and *common-base* configurations?

CHAPTER 15

Bipolar Buffer
Frequency-Domain Analysis

OBJECTIVES

- To analyze the *common collector* (*emitter follower*) buffer in the AC Sweep mode.
- To determine the A, Z_{in}, Z_{out}, and linearity characteristics of a buffer.
- To perform parametric analysis on a model parameter.

DISCUSSION

Figure 15.1 shows a Thevenized voltage source (V_S) and a Thevenized load (R_L) directly connected by a wire. Because both R_S and R_L are the same, the input voltage and input power are divided between the two resistors. Therefore, the circuit has a voltage and power "gain" of 1/2.

To prevent such losses, we interpose an active *buffer* between source and load, as shown in Figure 15.2. If the buffer has a Z_{in} of 40kΩ and a Z_{out} of 40Ω, the losses are nearly eliminated. As an added bonus, the buffer gives a positive power gain.

A *buffer*, therefore, is first and foremost an *isolation* circuit. It follows that its most important characteristics are high Z_{in} and low Z_{out}. Additional properties usually found with a buffer are a voltage gain near unity, high power gain, and a high degree of linearity.

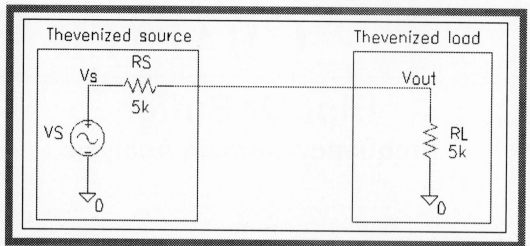

FIGURE 15.1

Direct connection
causes losses

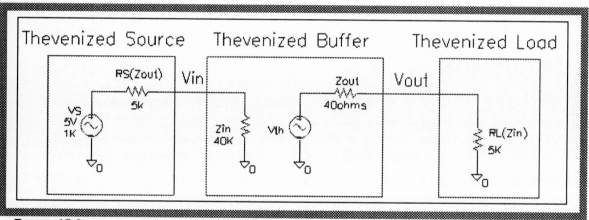

FIGURE 15.2

Thevenized buffer
circuit

One popular circuit that achieves all the characteristics of a buffer is the *common-collector* (grounded collector) configuration of Figure 15.3. Note that the output is taken off the emitter rather than the collector. Because the output voltage is nearly equal to the input voltage, the circuit is also known as an *emitter follower* (the emitter voltage follows the base voltage). In short, the output is approximately equal to the input in both amplitude and phase.

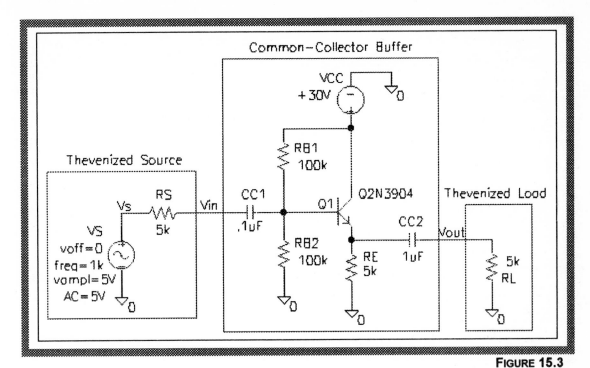

FIGURE 15.3

Common-collector
buffer

THEORY

We begin by calculating the approximate values for the buffer's AC characteristics. (Approximate because all capacitors and inductors take on their ideal values.)

$$re' = 25mV / I_{EQ} = 25mV / 3mA = 8.3\Omega$$

- $A(buffer) = V_{out} / V_{in} = (i_e \times R_E\|R_L) / (i_e \times (R_L\|R_E + re'))$

  $= 5k\Omega\|5k\Omega / (5k\Omega\|5k\Omega + 8.3) = .997 = -.026dB$

- $Z_{in}(buffer) = R_{B1}\|R_{B2}\|B(re' + R_E\|R_L) =$

  $100k\Omega\|100k\Omega\|175(8.3 + 5k\Omega\|5k\Omega) = 45k\Omega$

- $Z_{out}(buffer) = ((R_S\|R_{B1}\|R_{B2})/B + re')\|R_E =$

  $((5k\Omega\|100k\Omega\|100k\Omega)/175 + 8.3)\|5k\Omega = 35\Omega$

- G (power gain) $= \dfrac{.5Vout^2 / RL}{.5Vin^2 / Zin} \cong Z_{in}/R_L = 45k\Omega/5k\Omega = 9 = 19.1dB$

Looking at the gain equation [A(buffer)] we see that the linear term $(5k\Omega\|5k\Omega)$ is much larger than the nonlinear term (8.3Ω), and therefore we predict a highly linear output waveform.

FREQUENCY-DOMAIN ANALYSIS

In the previous chapter, we analyzed the small-signal amplifier completely in the time domain using transient analysis. In this chapter, we find that there are also many advantages to including the frequency mode. A major benefit is the ability to analyze characteristics over a range of frequencies.

Unless you are fortunate enough to have the right laboratory equipment (such as a *spectrum analyzer*), the activities of this chapter are best carried out under PSpice.

Remember, when using AC analysis, the default value for all variables is polar form magnitude (peak values).

SIMULATION PRACTICE

1. Draw the circuit of Figure 15.3 and set all attributes as shown.   (It is not necessary to draw the three boxes around the three stages.) Note that we have added an AC attribute (*AC = 5V*) to voltage source (V$_S$).

> Be aware that voltage source VSIN provides for all three types of analysis: DC, AC, and transient.

2. Set up the transient analysis for 0 to 2ms, and set up the AC analysis as follows:

    *   AC sweep type: Decade
    *   sweep parameters: Pts/Decade: 20
                            Start Freq.: 10
                            End Freq.: 100MEG

Voltage Gain, Power Gain, and Bandwidth

3. Use frequency (AC sweep) analysis to generate the amplitude plot of Figure 15.4.

4. Viewing the results (Figure 15.4), give an approximate value for the midband voltage gain. (*Midband* refers to the flat portion of the curve at the middle frequencies.)

        Midband voltage gain = _____

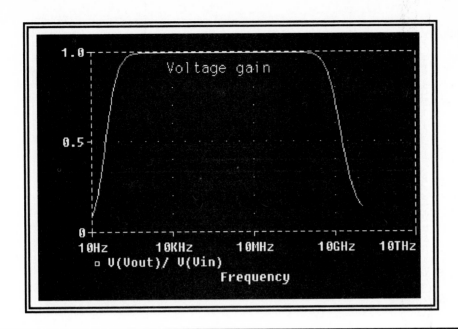

FIGURE 15.4

Buffer voltage
gain plot

5. *Bandwidth* refers to the range of frequencies where the gain is greatest. Specifically, bandwidth is the difference between the low- and high-frequency points where the gain drops to 70.7% of its peak midband value. Using this rule, what is the approximate bandwidth of the buffer?

   Bandwidth = _____

6. To determine power gain, add a plot of $[V_{out} \times I(R_L)]/[V_{in} \times I(R_S)]$. The result is shown in Figure 15.5.

7. What is the approximate midband power gain?

   Midband power gain = _____

8. Add a plot of phase to your graph—as shown in Figure 15.6.   ("d" = degrees.)

9. What is the approximate midband phase shift between input and output?

   Phase shift = _____

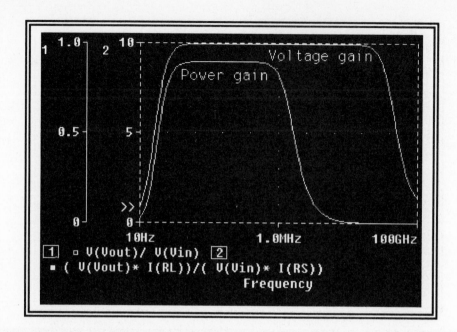

FIGURE 15.5

Adding power
gain

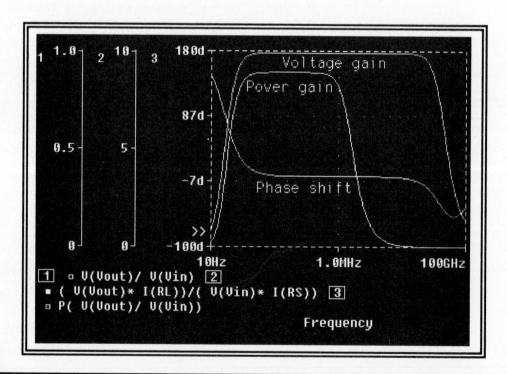

FIGURE 15.6

Adding phase

Z_{in} and Z_{out}

10. Because three Y-axes are the maximum allowed, delete one or all of the present curves.

    To determine Z_{in}, add a plot of $V_{in}/I(R_S)$—as shown in Figure 15.7. [Remember, the system automatically uses peak values for V_{in} and $I(R_S)$.]

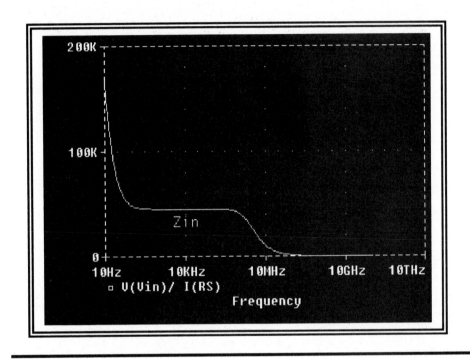

FIGURE 15.7

Plot of Z_{in}

11. What is the approximate midband Z_{in}?

    Z_{in} (midband) = _____

12. To measure Z_{out}, we set up the circuit as shown in Figure 15.8.

13. By plotting $V(Ve)/I(Vzout)$, generate the impedance plot of Figure 15.9 and report the midband Z_{out}. (Using V_e instead of V_{out} excludes the impedance of CC2 from the results.)

    Z_{out} (midband) = _____

14. Compare the PSpice results for A, G, Z_{in}, and Z_{out} with the theoretical calculations of the discussion. Are they *approximately* the same (in the same "ballpark")?

    Yes   **No**

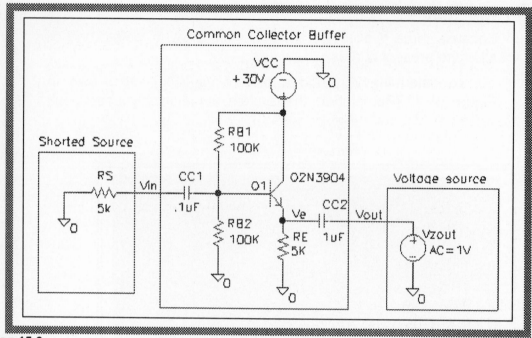

FIGURE 15.8

Determining
Z_{out}

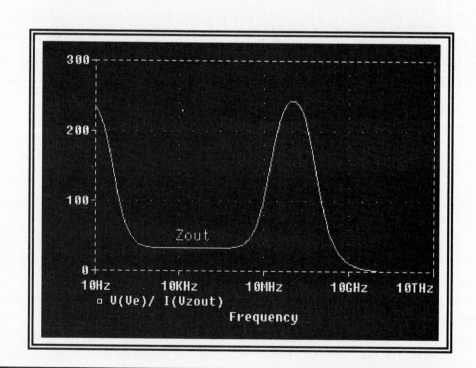

FIGURE 15.9

Plot of Z_{out}

PSpice for Windows

15. Return to the circuit of Figure 15.3 and select the transient mode. (Be sure the transient mode is enabled, set up, and all attributes are set as shown.) Generate V_{out} and determine the amplitude distortion from the following equation [frequency (harmonic) distortion techniques are the subject of Chapter 23]:

$$\% \text{ distortion} = \frac{\textbf{Vpeak(difference)}}{\textbf{Vpeak(average)}} \times 100 = \underline{\hspace{2cm}}$$

16. Based on all the previous results, does the circuit of Figure 15.3 have all the characteristics of a buffer?

     Yes   **No**

Bode Plot

17. When reporting gain as a function of frequency, it is common practice to switch from a linear Y-axis (Figure 15.4) to a logarithmic Y-axis (Figure 15.10).   (Remember, the X-axis is already logarithmic.) Such a log-log plot is called a *Bode* plot. The Y-axis values are transformed using the following equation— which is accomplished automatically by using the PSpice *dB* (or *DB*) operator.

     A(dB) = 20×logA(reg)

     Use the dB operator to generate the plot of Figure 15.10.

18. When switching to dB, 70.7% translates to –3dB.   Using the two frequency points where the gain drops to –3dB, again measure the bandwidth.   (How do the results compare with step 5?)

     Bandwidth = _____

Parametric Analysis of Model Parameters

As we saw from the theoretical calculations in the discussion, the input impedance (Z_{in}) is highly dependent on the value of *Beta*. One way to measure this dependency is to generate a family of curves for a variety of *Beta* values. In the 3904 model, parameter *Bf* is the *ideal maximum forward Beta*.

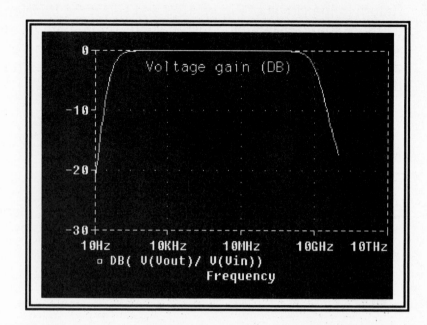

FIGURE 15.10

Voltage gain
in decibels

19. For the buffer circuit of 15.3, bring up the parametric dialog box (**Analysis**, **Setup**, enable the parametric mode, **Parametric**), fill in as listed below, **OK**, **Close**. (The AC analysis parameters remain the same.)

- Sweep Var. Type: **Model Parameter**
- Sweep Type: **Linear**
- Model Type: **NPN**
- Model Name: **Q2N3904-X**
- Param. Name: **Bf**
- Start Value: **100**
- End Value: **1000**
- Increment: **100**

20. Run PSpice and generate the Z_{in} family of curves of Figure 15.11.

21. Based on the results of Figure 15.11, by what approximate percentage did Z_{in} change when *maximum Beta* increased from 100 to 1000?

     % change in Z_{in} = _____

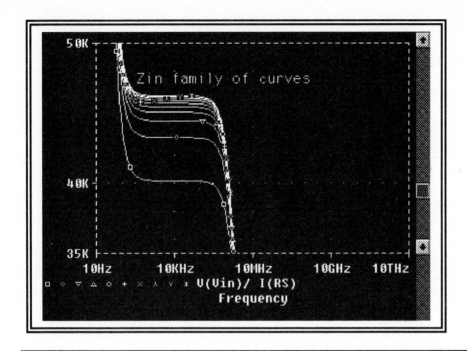

FIGURE 15.11

Family of Z_{in}
curves for
various B
(Y-axis adjusted)

The Darlington Buffer

22. Draw the buffer circuit of Figure 15.12, which uses a Darlington pair to greatly increase its "effective" B ($B_{total} = B1 \times B2$).

23. Using AC analysis, measure midband Z_{in} and Z_{out} for the Darlington buffer and compare to the non-Darlington values of steps 11 and 13.

Darlington	Non-Darlington
Z_{in} = _____	Z_{in} = _____
Z_{out} = _____	Z_{out} = _____

24. Based on step 23, does a Darlington buffer increase the isolation properties of a buffer?

        Yes   No

Advanced Activities

25. For the Darlington pair buffer of Figure 15.12, determine the amount of quiescent (DC) power dissipated in each transistor (Q1 and Q2). Which transistor is likely to require a heat sink?

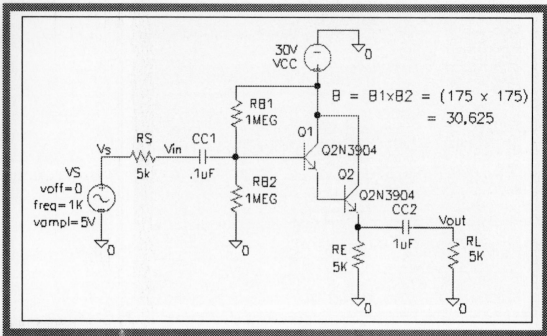

FIGURE 15.12

The Darlington
buffer

26. Verify that the coupling capacitors (CC1 and CC2 of Figure 15.3 or 15.12) are properly designed for a lowest-frequency-of-interest of 1kHz.

EXERCISES

- Perform a complete analysis of the amplifier/buffer of Figure 15.13, including such items as A (overall), G, Z_{in}, Z_{out}, and distortion. (How does the buffer stage "protect" the gain of the amplifier stage?)

- Investigate the properties of the *buffered voltage regulator* of Figure 15.14. (How low can we take R_L before the system comes out of regulation? Compare your answer to the results of Chapter 9.)

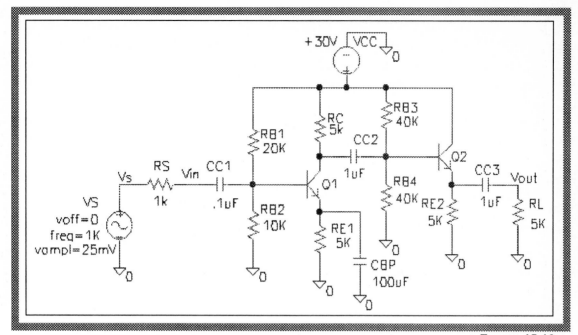

FIGURE 15.13

Amplifier/buffer

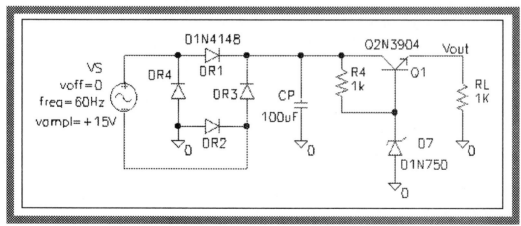

FIGURE 15.14

Buffered voltage
regulator

QUESTIONS & PROBLEMS

1. By using the words "high," "low," or "medium" in the following spaces, contrast the amplifier of Figure 14.1 with the buffer of this chapter.

	Amplifier	**Buffer**
A	_____	_____
Z_{in}	_____	_____
Z_{out}	_____	_____
Linearity	_____	_____

2. In the following statement, fill in each blank with "collector" or "emitter":

   To achieve voltage gain, we tap the output voltage of the _____, and to achieve buffering action, we tap the output voltage of the _____.

3. Why are the buffer circuits of this experiment called *emitter followers*? (What is the phase relationship between input and output voltage?)

4. Why are all the buffer circuits of this experiment highly linear? Would they remain highly linear if the load were reduced to 25Ω? Why?

5. For every 10,000 electrons that enter the emitter of a Darlington pair, how many electrons leave the base?   (Assume $B = 175$ for each transistor.)

6. Referring to Figure 15.2, for maximum transfer of voltage from left to right (source to load), which of the following should exist?:

   (a) Z_{in} low, Z_{out} low
   (b) Z_{in} low, Z_{out} high
   (c) Z_{in} high, Z_{out} low
   (d) Z_{in} high, Z_{out} high

7. Referring to Figure 15.15, why is the input impedance at the base equal to 100kΩ.   Why is the output impedance at the emitter 10Ω? (Assume that re' is zero.)

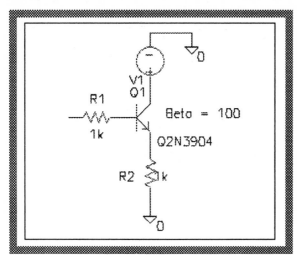

FIGURE 15.15 Transistor circuit

8. Referring to Figure 15.7, why does Z_{in} go up at low frequencies and down at high frequencies?

CHAPTER 16

Amplifier Power
Class A Operation

OBJECTIVES

- To design a Class A amplifier.
- To generate a load line.
- To determine power factors.

DISCUSSION

Unlike the small-signal amplifier of Chapter 14, the Class A amplifier of Figure 16.1 is designed for large voltage and current applications (note the small resistor values). Therefore, power is an important consideration. Because we have access to a split power supply, we choose emitter biasing.

In particular, our design must provide the following:

- The maximum possible unclipped output voltage.
- The ability to handle the heat dissipated in the transistor.

THE LOAD LINE

Our primary Class A design aid is known as a *load line*. As shown by Figure 16.2, we first locate the Q point, then we draw a line through the Q point whose inverse slope is equal to the total AC load.

When an AC signal is present, the operating point moves back and forth along the load line about the Q *point*. The two endpoints are called $I_{c(sat)}$ and $V_{ce(off)}$. If the circuit is well designed, the Q point is *centered* and we generate the largest possible output signal without clipping.

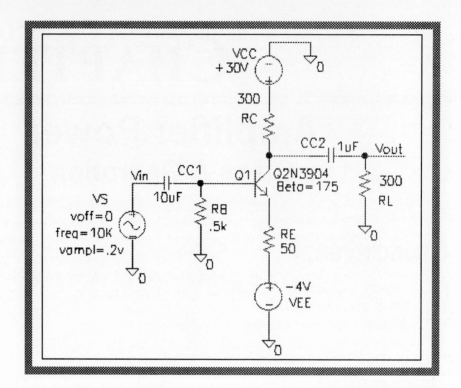

FIGURE 16.1

Initial Class A
amplifier design

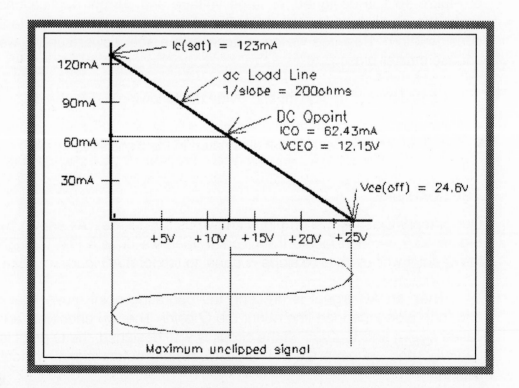

FIGURE 16.2

The load line

PSpice for Windows

The values shown on Figure 16.2 were calculated as listed below. (The saturation and cutoff values are easily determined from the geometry of similar triangles.)

- $I_{EQ} = I_{CQ} = (4V - .7V) / (50\Omega + .5k\Omega/175) = 62.43mA$

- $V_{CEQ} = V_{CQ} - V_{EQ} = (30V - 62.43mA \times 300) - (-4V + 62.43mA \times 50)$
$$= 11.27V - (-.88V) = 12.15V$$

- rl (ac load) $= R_C\|R_L + R_E = 300\Omega\|300\Omega + 50\Omega = 200\Omega$

Based on Figure 16.2, we predict that the circuit is well designed (has a nearly centered Q point). We further predict that its *compliance* (maximum unclipped signal) will be approximately 12.15V. (However, this is not the compliance of the "true" output signal across R_L because R_E absorbs 25% of the load voltage. Therefore, we predict that the output voltage compliance across R_L is 75% of 12.15V, or 9.11V.)

POWER CONSIDERATIONS

- The *load power* (P_L) is the average AC power developed across resistor R_L (the "true" load) when driven at its maximum unclipped level. For a compliance of 9.11V:

$$P_L = .5 \times V_{out}^2 / R_L = .5 \times 9.11^2 / 300\Omega = 138mW$$

- The *source power* (P_S) is the average power supplied by the DC power supplies.

$$P_S = (V_{CC}+V_{EE}) \times I_C(average) = (30V + 4V) \times 62.43mA = 2.1W$$

- The *dissipated power* (P_D) is the average power deposited into the transistor. Because *AC* voltage and current are out of phase within the transistor, the "worst case" dissipated power is the quiescent *DC* power.

$$P_D = V_{CEQ} \times I_{CQ} = 12.4V \times 62.43mA = 774mW$$

- The *efficiency* (η) is determined by dividing the maximum load power by the source power.

$$\eta = P_L / P_S \times 100 = 138mW / 2.1W = 6.6\%$$

In this chapter, we verify all of these predictions using PSpice. During our experimental activities, a major consideration will be: Should we use *transient* or *AC* analysis?

SIMULATION PRACTICE

1. Draw the circuit of Figure 16.1. (<u>Note</u>: If V_{EE} were turned upside-down, it's voltage attribute would be +4V.)

2. To determine the maximum unclipped output signal, we select transient analysis. We will input a sine wave of gradually increasing magnitude and note when clipping occurs on the output. To generate a "reverse damped" sine wave, first review *Schematic Note 16.1*.

Schematics Note 16.1
How do I generate a damped sine wave?

VSIN uses the following formula to generate a sine wave:

$$V_{out} = V_{off} + V_{ampl} \times \sin\{2\pi \times [freq \times (time - td) + phase/360]\} \times e^{-(time - td) \times df}$$

A "true" damped sine wave (amplitude decreases with time) is created from positive values of df (*damping factor*), and a "reversed" damped sine wave (increases with time) is created from negative values of df. To generate a damped sine wave, bring up VSIN's *Part Name* dialog box (**DCLICKL** on VSIN symbol) and enter df along with the usual parameters.

As an example, let's create an increasing sine wave that starts from 200mV and rises to approximately 10V in 10 cycles of a 10kHz waveform. For this case, df is calculated as follows:

$$e^{-(1ms \times df)} = 10V/.2V = 50 \qquad\qquad df \cong -4000$$

---

3. Using a "reverse damped" sine wave, generate the input/output waveforms of Figure 16.3. (<u>Hint</u>: Use V_{ampl} = .2V and df = −4000, as determined in *Schematics Note 16.1*.)

4. Do the waveforms of Figure 16.3 verify that the circuit is well designed? (Does the output signal clip on both ends at approximately the same time, indicating the Q point is approximately centered?)

   Yes   **No**

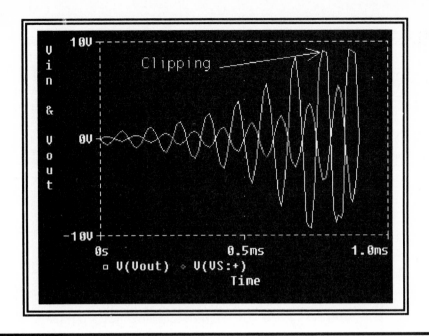

FIGURE 16.3

Looking for clipping

5. Is the output signal compliance (maximum unclipped output signal) approximately 9V, as predicted?

    Yes   **No**

Power

Because we are interested in power over a range of input signal *amplitudes*, we remain with a transient analysis.

6. To perform a power analysis, generate the *instantaneous* power waveforms of Figure 16.4.

7. Using results at the maximum unclipped signal point, determine each of the following: [Suggestion: Open a second window and regenerate the power curves using the AVG (running average) operator.]:

   • P_L **(average) =** _____

   • P_S **(average) =** _____

   • P_D **(average) =** _____

   • η **=** _____

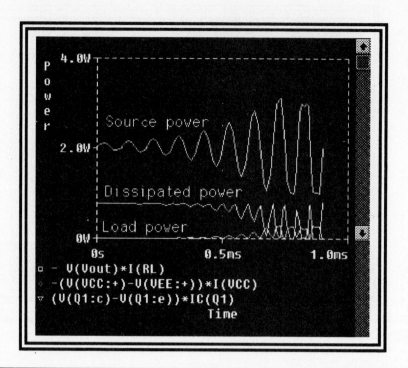

FIGURE 16.4

Power graphs

8. Do the experimental values of step 7 approximately equal the theoretical values calculated in the discussion?

   Yes **No**

9. Using frequency domain analysis, determine the bandwidth of the amplifier of Figure 16.1. (<u>Hint</u>: Add attribute *AC=* to V_S and setup the *AC Sweep* mode.)

   Bandwidth = _____

Advanced Activities

10. Generate the "real-time" load line of Figure 16.5. (<u>Hint</u>: Start with the same damped sine wave Probe data as the other plots, but be sure to redefine the X-axis.) Can you tell approximately where the Q point is by the thickness and intensity of the curve? Compare the result to the predicted load line of Figure 16.2.

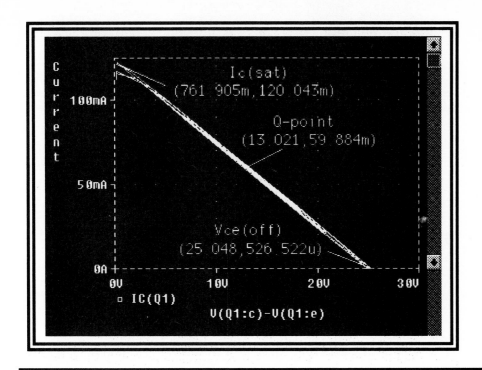

FIGURE 16.5

Load line created
by PSpice

EXERCISES

- Check the audio amplifier of Chapter 15 (Figure 15.13) for **Class A** operation. If necessary, make changes to bring it into **Class A** operation.

QUESTIONS & PROBLEMS

1. How does a Class A amplifier differ from a small-signal amplifier?

2. If the Q point is centered, the output signal clips

    (a) at the positive peaks first.
    (b) at the negative peaks first
    (c) at both positive and negative peaks at the same time.

PSpice for Windows

3. Would you say that the efficiency of a Class A amplifier is high or low?

4. For a sine wave across the load resistor (R_L), the average current and voltage are zero.   Why is the average *power* not zero?

5. Why does the worst case dissipated power (power lost in the transistor) occur when the *AC* signal input is zero?

6. Why do *power* amplifiers use smaller resistors than *small-signal* amplifiers?

7. Looking at the load line of Figure 16.5, why is the thickness and intensity of the load line greatest about the Q point?

8. What is the difference between *power* and *energy*?

9. Why is a load line straight (linear)?

10. If the load (R_L) is increased, would the load line slope of Figure 16.5 increase or decrease?

CHAPTER 17

Amplifier Efficiency
Classes B & C

OBJECTIVES

- To design and analyze class B and C amplifiers.
- To compare the efficiency of Class A, B, and C amplifiers.

DISCUSSION

The Class A amplifier of Chapter 16 suffers from notoriously poor efficiency. To increase efficiency greatly, we switch to the Class B and C amplifier/buffer designs of this chapter.

- The Class B circuit of Figure 17.1 employs two transistors in a "push/pull" configuration. Because we have the luxury of a split power supply, no coupling capacitors are required. It is called Class B because each transistor is on (conducts) for approximately 50% of each cycle. The upper (NPN) transistor conducts during the positive half cycles and the lower (PNP) conducts during the negative half cycles.

    The Class B configuration is usually employed as a buffer and power amplifier in the common-collector (emitter-follower) configuration. A Class B buffer can yield more than 75% efficiency because very little power is wasted in biasing the circuit.

- The Class C circuit of Figure 17.2 is the most efficient of all. When in operation, it simulates the action of a hammer and a bell. A biased clipper is the hammer, and a tank circuit is the bell.

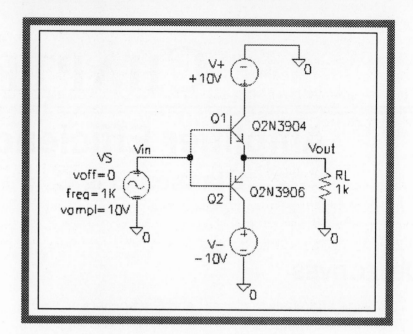

FIGURE 17.1

Class B operation

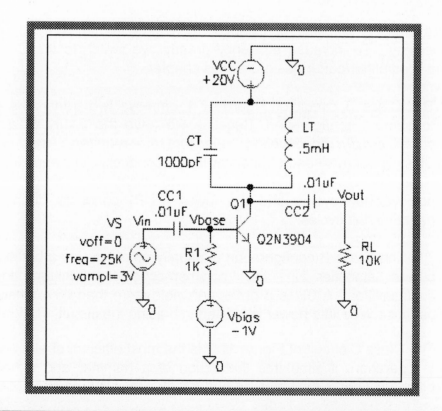

FIGURE 17.2

Class C operation

PSpice for Windows

At the top of each input cycle, the transistor saturates, the collector is grounded, and the capacitor is suddenly charged (the hammer hits the bell). The transistor then goes into cutoff for the rest of the cycle and the capacitor and inductor trade energy (the bell rings). It is called Class C because the transistor is on for much less than 50% of each cycle.

A Class C amplifier typically yields more than 90% efficiency because almost no power is lost in biasing the circuit or dissipated in the transistor. Because of its high efficiency and the use of a resonant tank circuit, the Class C amplifier is typically used as a common-emitter radio-frequency amplifier for frequencies above 1MHz .

SIMULATION PRACTICE

Class B Buffer

1. Draw the Class B buffer of Figure 17.1 and set the attributes as shown.

2. Using PSpice, generate the transient output waveform and determine the voltage gain.

   Class B voltage gain (A) = _____

   Is the voltage gain what you would expect of a buffer?

   Yes   No

3. Zoom in (**Zoom**, **Area**) on the output waveform at the point where it crosses the 0V axis. (Use the steady-state 1.5ms point, as shown in Figure 17.3). This is called *cross-over distortion*, and is caused by the barrier potential of the two transistors. (While the input signal is between plus and minus .7V, the output signal is zero.)

4. To overcome cross-over distortion, add the *trickle-bias* circuit of Figure 17.4. The circuit makes use of the similar barrier potential characteristics of diodes to bias each transistor just beyond its knee.

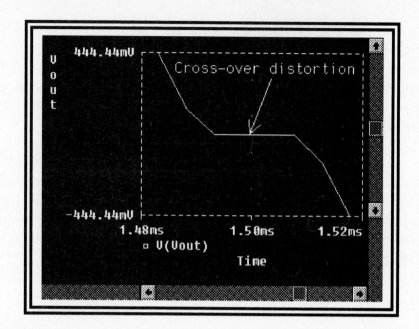

FIGURE 17.3

Crossover distortion

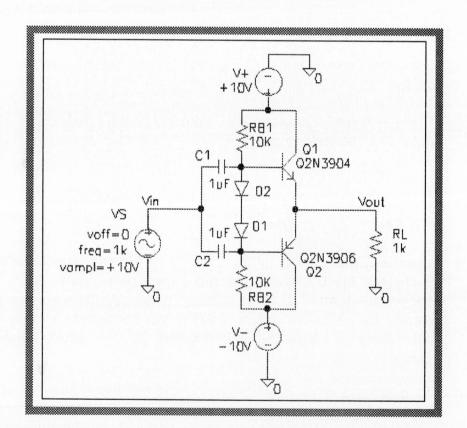

FIGURE 17.4

Adding trickle bias

5. Generate new output waveforms. Is the cross-over distortion gone?

    Yes   No

Efficiency

6. Efficiency is equal to *average* load power divided by *average* source power (reported as a percentage).

    For the trickle bias circuit of Figure 17.4, use the following equations to generate the *instantaneous* load and source power curves of Figure 17.5. (The minus signs are necessary because source and emitter currents are negative.)

    - $P(load) = -V(V_{out}) \times I(R_L)$
    - $P(source) = -(V(V+:+) \times I(V+) + V(V-:+) \times I(V-))$

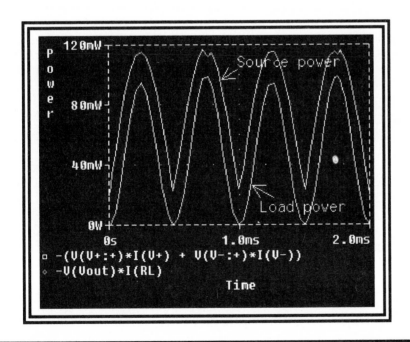

FIGURE 17.5

Class B instantaneous power curves

7. From the curves of Figure 17.5, we could obtain average power through mathematical analysis. Instead, use the AVG (average) operator to generate the running average curves of Figure 17.6.

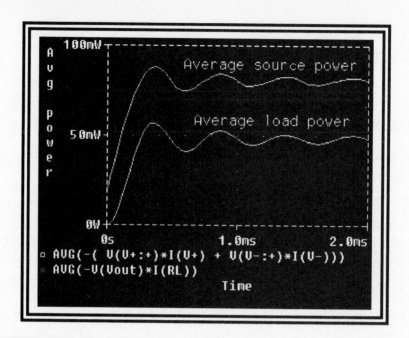

FIGURE 17.6

Class B average
power curves

8. Use your cursor to determine the steady-state average source and load power (the values near the right-hand side of the graph).

> **Average load power**   = _____
>
> **Average source power** = _____

9. From the results of step 8, determine experimental efficiency. (How does it compare with the maximum value mentioned in the discussion?)

$$\text{Efficiency} = \frac{\textbf{Average load power}}{\textbf{Average source power}} \times 100 = \underline{\hspace{2cm}}$$

Class C Amplifier

10. Draw the circuit of Figure 17.2 and set the attributes as shown. (The frequency value of V_S will be set later.)

11. Determine the resonant frequency of the tank circuit.

$$F(\text{resonant}) = \frac{1}{2\pi \times (L \times C)^{1/2}} = \underline{\hspace{2cm}}$$

12. Set the frequency value of V$_S$ to approximately 1/10 of the resonant frequency.

13. Generate the input, base voltage, and output waveforms of Figure 17.7.

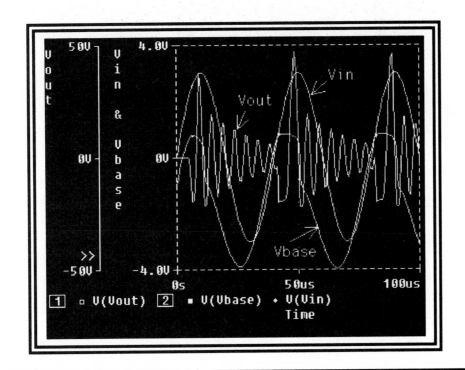

FIGURE 17.7

Class C input and output waveforms

14. (a) Is the base voltage negatively clamped?

    Yes   **No**

    (b) Is the output waveform a damped sine wave at the resonant frequency?

    Yes   **No**

15. As time permits, change the value of the input frequency and note the result.

Advanced Activities

16. Perform an energy analysis of the Class C amplifier and *estimate* its efficiency.

EXERCISES

- Using PSpice results, compare the overall efficiency of the audio amplifier of Figure 17.8 (using a Class B output stage) with the overall efficiency of the audio amplifier of Figure 15.13 (using a conventional output stage).   Note that the amplifier uses a single power supply.

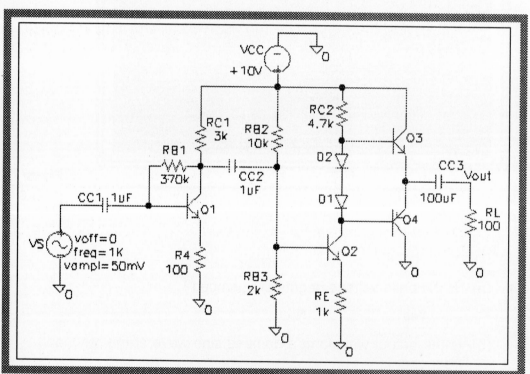

FIGURE 17.8

Audio amplifier

QUESTIONS & PROBLEMS

1. Referring to the biased Class B amplifier stage of Figure 17.8, what powers the circuit during the half cycle when the upper NPN transistor is biased off? (Hint: What circuit component stores energy?)

2. Why does the trickle-bias circuit of Figure 17.4 consume very little power?

3. Assuming that the transistors of Figure 17.1 have a *Beta* of 175, what is the approximate power gain?

4. The input circuit to the base of a Class C amplifier is

    (a) a positive clamper.
    (b) a negative clamper.

5. Viewing the Class C waveforms of Figure 17.7, what causes the output waveform to decay between "hits"?

6. Why is the efficiency of a Class C amplifier so high? (Hint: Why does most of the source energy pass on to R_L?)

7. For each of the amplifier classes listed below, approximately what percentage of the time is a given transistor on?

    Class A _____

    Class B _____

    Class C _____

8. Why is a Class C amplifier generally reserved for high (radio) frequencies?

9. Why is the trickle bias circuit of Figure 17.4 often called a *current mirror*?

PART IV
The Field-Effect Transistor

In Part IV, we move from the Bipolar to the field-effect transistor (FET). As with Part III, we concentrate primarily on amplifiers and buffers.

We will see how the FET's inherent high input impedance and vastly different transconductance properties influence its characteristics.

CHAPTER 18

Field-Effect Transistor Characteristics
Drain Curves

OBJECTIVES

- To display FET "master" and "slave" curves.
- To determine a FET's input and output impedance.
- To determine the effects of temperature on a FET.

DISCUSSION

Like the bipolar transistor, a *field-effect transistor* (FET) is also a *voltage-controlled current source* (VCIS). That is, an input voltage controls an output current, regardless of the output voltage. However, the way the FET achieves its VCIS characteristics is vastly different.

The FET comes in a number of different types. The JFET (*junction field-effect transistor*) of Figure 18.1 was the first developed. It achieves its VCIS characteristics as follows: As the positive drain/source voltage increases it "pulls" on electrons, creating a drain current in the N region. However, because of the reverse-biased PN junction between gate and drain, it also creates a positively charged *depletion region* that "squeezes" the drain current.

In normal operation, when the *drain voltage* is above a certain threshold (*pinchoff*), this "pulling" and "squeezing" cancel—*and the drain becomes a current source* (current independent of voltage). However, the *input* gate/source voltage is not subject to cancellation (it "squeezes" only) and therefore *does* control the current.

In summary, the input gate/source ("master") voltage controls the output drain ("slave") current, regardless of the drain/source voltage—making the FET a *voltage-controlled current source* (VCIS).

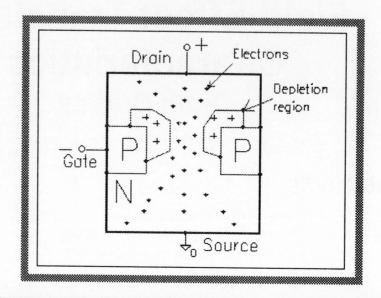

FIGURE 18.1

JFET internal
operation

THE BIPOLAR VERSUS THE JFET

If both the bipolar transistor and FET are simply VCISs, then how do they differ? The most important differences are the JFET's extremely high input impedance and its vastly different input/output (transconductance) characteristics. These two differences give the JFET an edge when used in certain applications.

TEST CIRCUIT

The test circuit of Figure 18.2 shows an N-channel JFET. The JFET also comes in the P-channel version, but may not be available in the PSpice evaluation library (*eval.lib*). The P-channel is identical in operation to the N-channel, except all voltages and currents are reversed.

TRANSCONDUCTANCE

When operated as a VCIS, the output of a FET is drain current (I_D), and the input is gate/source voltage (V_{GS}). This ratio ($\Delta I_D / \Delta V_{DS}$) is called the transistor's *transconductance* (g_m) and is given in units of μSeimens (μS). A typical value for transconductance is 4000μS (or 1/250Ω).

As before, one of the best ways of investigating the VCIS characteristics of a JFET is by generating drain and gate/source curves. This is done with the test circuit of Figure 18.2.

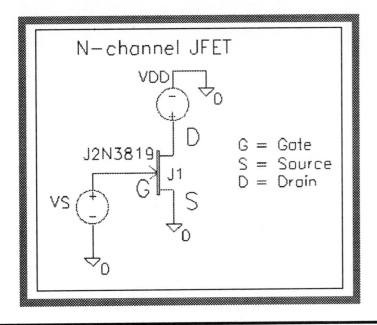

FIGURE 18.2

N-channel JFET
test circuit

SIMULATION PRACTICE

1. Draw the circuit of Figure 18.2 and set the attributes as shown. (V_S and V_{DD} will be swept and therefore need not be assigned bias point values at this time.)

Drain curves

2. Generate the set of FET drain curves of Figure 18.3. (<u>Hint</u>: V_{DD} is the main sweep and V_S is the nested sweep.)

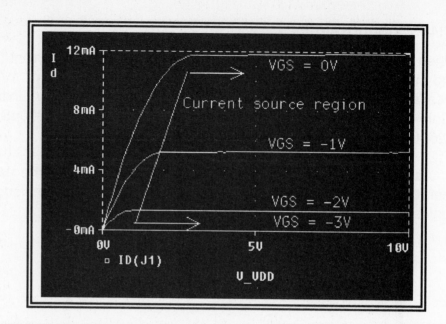

FIGURE 18.3

Drain curves for
JFET

3. By taking measurements, answer the following:

    (a) What is the maximum possible current (known as I_{DSS})? What input voltage (V_{GS}) produces I_{DSS}?

    I_{DSS} = _____ V_{GS} @ I_{DSS} = _____

    (b) What gate voltage causes *pinchoff* ($I_D = 0$)?   This value is called V_{GSoff}

    V_{GSoff} = _____

    (c) Determine a typical value for output (drain) impedance. (<u>Hint</u>: Measure 1/slope for a typical point in the current source region, or use the "d" operator.)

    Z_{out} = _____

The Gate/Source Curve

4. By taking values from Figure 18.3 (all in the current-source region), draw a *transconductance* curve on the graph of Figure 18.4.

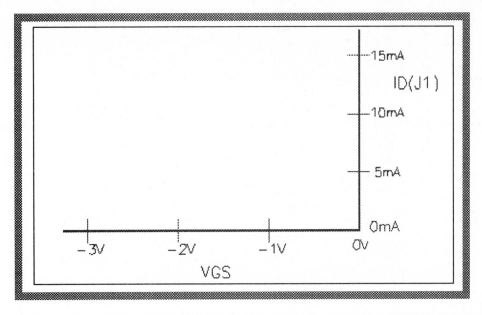

FIGURE 18.4

Transconductance
curve

5. Using PSpice, generate the transconductance curve of Figure
    18.5 *directly*, and compare to the *indirect* method of Figure 18.4.
    (<u>Hint</u>: Set V_{DD} to +10V and do a DC sweep of V_S.)

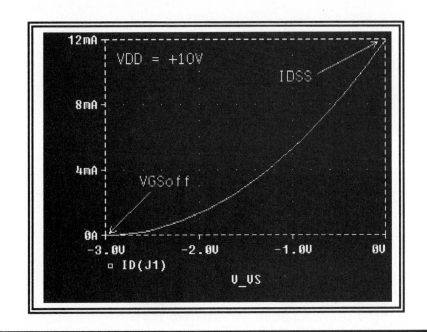

FIGURE 18.5

JFET
transconductance
curve

PSpice for Windows

6. The "master" curve of Figure 18.5 is a *parabola* that obeys the following equation:

    $$I_D = I_{DSS} (1 - V_{GS}/V_{GSoff})^2$$

    As an example of the use of this equation, at what V_{GS} does $I_D = 1/2\ I_{DSS}$ (a good Q point location)? Does your value agree with Figure 18.5?

    $$V_{GS}\ (@1/2I_{DSS}) = \text{_____}$$

7. Transconductance is the *slope* of the curve (where slope = $\Delta I_D/\Delta V_{GS}$). Using the differentiate ("d") operator, add a curve of transconductance to the graph of Figure 18.5. (The result is shown in Figure 18.6.)

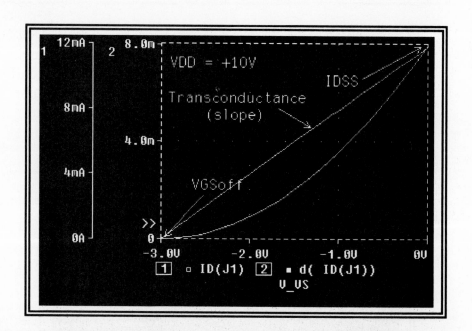

FIGURE 18.6

Adding a graph
of slope

8. Based on the results of Figure 18.6, what is the transconductance at the "half-current" Q point (where $I_D = 1/2\ I_{DSS}$)?

    Transconductance @ half-current (μS) = _____

9. Add a 3rd Y-axis to the graph of Figure 18.6 and plot input impedance   [V(j1:g)/IG(J1)].   Record below a typical value for Z_{in}.   (*Note*: "T" = *tera* = 10^{+12}.)

> Z_{in} **(typical) =** _____

Based on your results, would you say that a JFET has a naturally high input impedance?

> **Yes   No**

"Master" Curve Temperature Effects

10. To determine how temperature affects the "master" (transconductance) curve of Figure 18.5, recreate the graph of Figure 18.7.   (Hint: Make temperature a nested DC sweep from -50°C to +50° in increments of 25°C.)

11. Based on the results of Step 10, answer the following: When the temperature increases from 0°C (32°F) to +25°C (103°F), IDSS changes by what percent?

> **% change in I_{DSS} =** _____

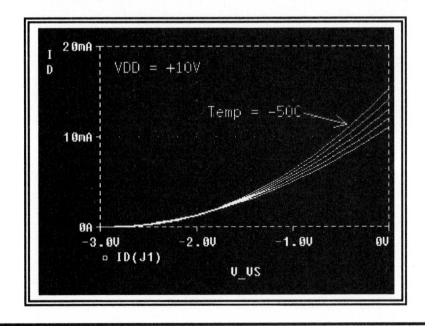

FIGURE 18.7

JFET temperature effects

Advanced Activities

12. Figure 18.8 shows the "master" curve for the irf150 power E-MOSFET (enhancement MOSFET). Draw a test circuit, generate a set of drain ("slave") curves, and sketch the results in Figure 18.9. (How does it differ from the JFET?)

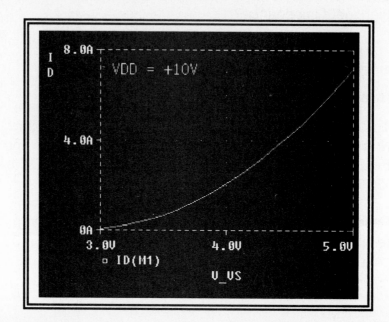

FIGURE 18.8

"Master" curve for
irf150 E-MOSFET

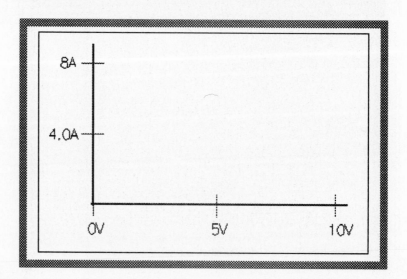

FIGURE 18.9

Drain ("slave") curves for
irf150 power MOSFET

EXERCISES

- Using the irf150 (N-channel E-MOSFET) and irf9140 (P-channel E-MOSFET), draw the digital CMOS (*complementary-symmetry metal-oxide semiconductor*) inverter of Figure 18.10(a) and generate the test waveforms of Figure 18.10(b).

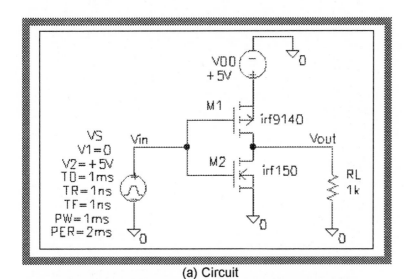

(a) Circuit

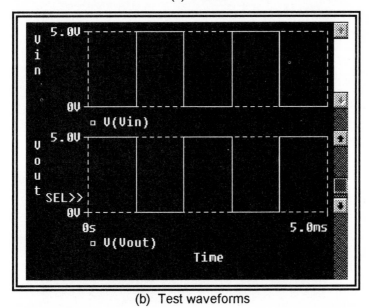

(b) Test waveforms

FIGURE 18.10

The CMOS inverter
(a) Circuit
(b) Test waveforms

- Looking at Figure 18.3, we see that the ohmic region (to the left of the current-source region) acts as a *voltage-variable resistor* for small voltages. Using the test circuit of Figure 18.11(a), generate the curves of Figure 18.11(b). Based on your results, create a table of Vcontrol versus Rfet. (<u>Hint</u>: Use a Vcontrol variable list of 0, -2.8, -2.9, -2.95, -3.)

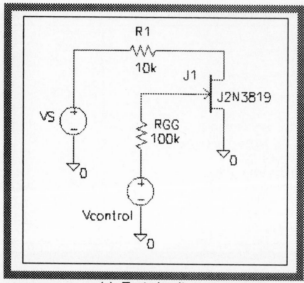

(a) Test circuit

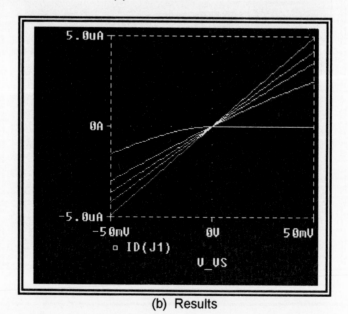

(b) Results

FIGURE 18.11

Voltage-variable resistor
(a) Test circuit
(b) Results

PSpice for Windows

QUESTIONS & PROBLEMS

1. Why is the input impedance of a JFET so high?

2. Why is a JFET (when biased as in Figure 18.1) a VCIS (rather than a ICIS)?

3. Which of the following typically has the highest *transconductance* (output current divided by input voltage)?:

   (a) Bipolar transistor
   (b) FET

4. Write "bipolar" or "JFET" before each statement.

   _____ Normally off, requires voltage to turn on.

   _____ Normally on, requires voltage to turn off.

5. The transconductance of a JFET is highest

   (a) at low values of drain current.
   (b) at high values of drain current.

6. What do the NPN transistor and E-MOSFET have in common?   (<u>Hint</u>: compare the "master" curves of each.)

7. A transconductance of 4000µS is equivalent to what in inverse ohms?

CHAPTER 19

FET Biasing
Stability

OBJECTIVES

- To design and analyze several popular FET biasing circuits.
- To determine how temperature affects stability.

DISCUSSION

We bias a JFET for the same reason we bias a bipolar transistor: to place its quiescent (Q) point at an appropriate place in the "master" curve.

Based on the graph of Figure 19.1 (reproduced from Chapter 18), this requires a negative DC voltage across its gate/source junction. When properly biased, the superimposed AC signal will have room to operate on both its positive and negative cycles.

As with the bipolar transistor, we can choose from a variety of JFET biasing circuits. The major considerations are simplicity, stability, and flexibility.

SIMULATION PRACTICE

Self-bias

1. The simplest type of JFET biasing is the *self-biased* circuit of Figure 19.2. (*Self* because rising current through R_S automatically places a positive bias voltage at the source, while the gate voltage remains zero.) Draw the circuit and set the attributes as shown. (The value of R_S is determined later.)

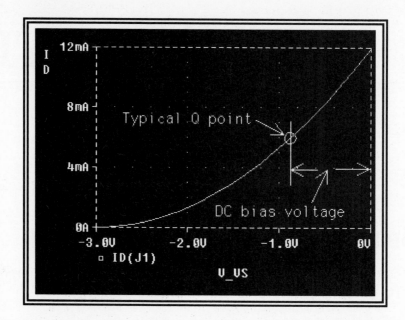

FIGURE 19.1

JFET biasing and
Q point

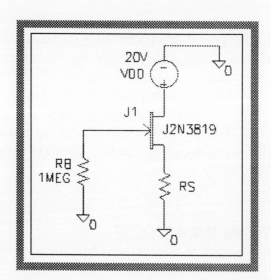

FIGURE 19.2

Self-biasing

2. By solving the following equations, determine the gate/source quiescent voltage (V_{GS}) that places the Q point at the "half-current" biasing point (when $I_D = 1/2\ I_{DSS}$).

I_{DSS} = 12mA   (from Figure 19.1)

V_{GSoff} = -3V   (from Figure 19.1)

$I_D = I_{DSS}\ (1 - V_{GS}/V_{GSoff})^2$

V_{GS} (for half-current biasing) = _____

3. Is the value for V_{GS} (@ half-current) calculated in step 2 close to that indicated by Figure 19.1?

   Yes   No

4. From the value of V_{GS} (@ half-current) determined in step 2, use Ohm's law to calculate the value of R_S required to give this "half-current" Q point.   (Hint: V_G is approximately zero.)

   R_S (half-current bias point) = _____

5. Set R_S to the value determined in step 4.   Using PSpice, perform a bias point analysis and record the Q point values below.   Are the actual values approximately equal to the expected values of step 2?

   I_{DQ} = _____ V_{GSQ} = _____

Temperature Effects

6. To determine temperature stability, we will see how the drain current [$I_{D(J1)}$] changes with temperature. Generate the graph of Figure 19.3 and determine the following:

   $\Delta I_D\ /\ \Delta Temp$ = _____

Voltage-Divider Bias

7. Because Figure 19.3 proves that self-bias is unstable, we turn to the *voltage-divider* bias circuit of Figure 19.4.

   Use Ohm's law to determine the value of R_S required to place the Q point at the half-current point of Figure 19.1. (Hint: Vsource will be more positive than Vgate by the DC bias voltage.)

   R_S (half-current point) = _____

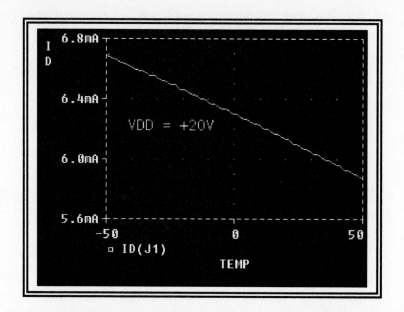

FIGURE 19.3

Self-bias temperature
stability.

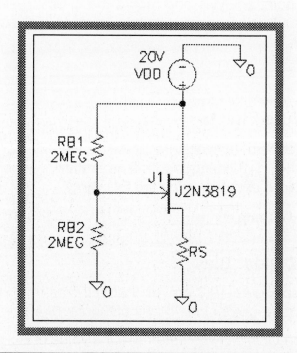

FIGURE 19.4

Voltage-divider bias

8. Set R_S to the value determined in step 7. Using PSpice, perform a bias point analysis and record the Q point values. Are the actual values approximately equal to the expected values?

   I_{DQ} = _____ V_{GSQ} = _____

9. Perform a temperature analysis of voltage-divider bias (similar to step 3) and report the following:

   ΔI_{DQ} / Δ**Temp** = _____

10. Comparing steps 6 and 9, give one reason why voltage-divider bias is superior to self-bias.

Advanced Activities

11. Figure 19.5 shows how the bias resistor (R_S) for the self-biased JFET circuit of Figure 19.2 can be determined graphically (for the "half-current" Q point). Use this same method to determine the "half-current" R_S value for the voltage-divider circuit of Figure 19.4. (<u>Hint</u>: Assume that the initial value of V_{GS} = +10V.)

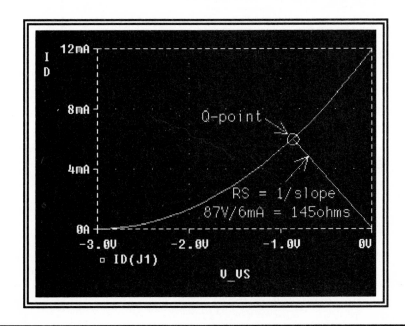

FIGURE 19.5

Graphical design

12. Based on the graphical data of Figure 18.8 (previous chapter), determine the value of R_D that will place the E-MOSFET biasing circuit of Figure 19.6 at the half-current Q point. Check your results using PSpice. (<u>Hint</u>: Assume that $I_{RG} = 0$.)

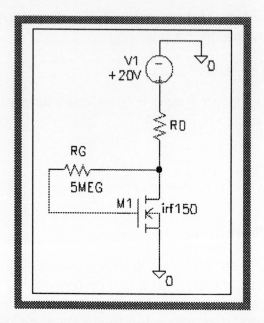

FIGURE 19.6

Drain feedback
biasing of E-MOSFET

EXERCISES

• Perform a power analysis of the CMOS inverter circuit of Figure 19.7 [reproduced from Figure 18.10(a)]. Based on the results, what is an important feature of CMOS digital circuits? During what times was power delivered by the source? Why would power drain go up as input frequency went up?

QUESTIONS & PROBLEMS

1. Referring to Figure 19.1, when the current is at the "half-current" point (6mA), the voltage is *approximately* at the

   (a) "quarter-voltage" point (−3/4V).
   (b) "half-voltage" point (−1 1/2V).

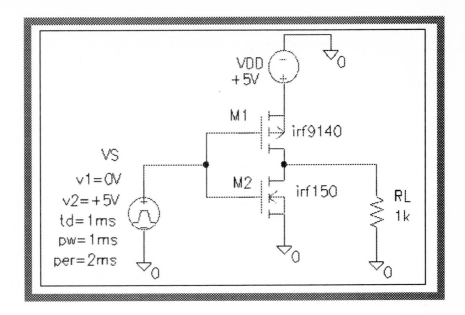

FIGURE 19.7

CMOS inverter

2. Why is the biasing technique of Figure 19.2 called "self-biasing"?

3. Based on the transconductance curve of Figure 19.1, what gate voltage will place the JFET shown below at the "half-current" point?

    V_G **(half-current) =** _____

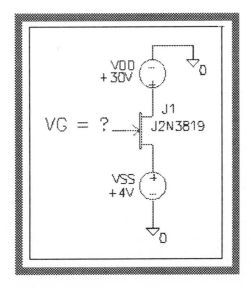

4. Referring to the voltage-divider biasing circuit of Figure 19.4, why can we be quite sure that the gate voltage is very close to 10V?

5. Can we use the drain feedback biasing method of Figure 19.6 on a JFET? Why or why not?   (<u>Hint</u>: With a JFET can V_G ever equal V_D?)

CHAPTER 20

FET Amplifiers & Buffers
Input Impedance

OBJECTIVES

- To design, analyze, and compare JFET amplifiers and buffers.
- To determine FET circuit input and output impedance.

DISCUSSION

Amplifiers and buffers made with JFETs are similar in design and function to those made with bipolar transistors. The major difference is that FET circuits offer a much higher input impedance, and the differences in transconductance characteristics lead to differences in gain and linearity.

SIMULATION PRACTICE

Amplifier

1. Draw the JFET amplifier of Figure 20.1 and set the attributes as shown. (Note that the amplifier is biased by the same half-current self-biasing circuit analyzed in the previous chapter.)

2. Using PSpice, run a bias point solution. Examine the output file, and mark the Q point on the transconductance curve of Figure 20.2. Is the Q point at the expected half-current point? (Figure 20.2 is a copy of Figure 18.6 and need not be generated by the student.)

    Yes   **No**

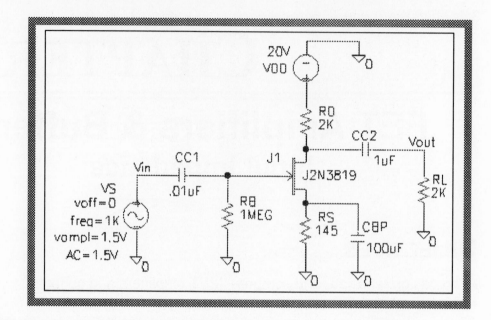

FIGURE 20.1

JFET amplifier
using self biasing

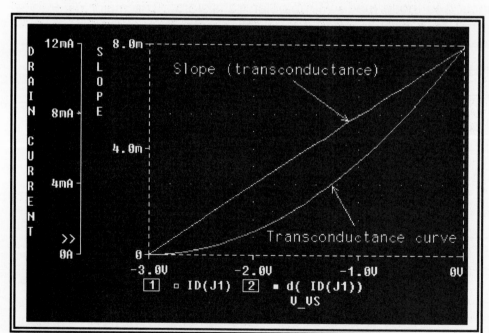

FIGURE 20.2

JFET
transconductance
curve

PSpice for Windows

3. *Transconductance* is the slope of the transconductance curve.

    Based on Figure 20.2, what is the transconductance (g_m) at the Q point?   [Hint: Make use of the slope curve generated by the differential (d) operator.]

    g_m (@ Q point) = _____ μSiemens

4. Using the value of g_m from step 3, calculate by hand the "big three." [Hint: As an approximation, assume the ideal value of infinity for both Z_{in}(FET) and Z_{out}(FET).]

    A $=$ $R_L\|R_D \times g_m$ $=$ _____

    Z_{in} $=$ $R_B\|Z_{in}$(FET) $=$ _____

    Z_{out} $=$ $R_D\|Z_{out}$(FET) $=$ _____

5. Using either the transient or AC mode, determine the "big three" using PSpice and compare to the theoretical results of step 4.

    A (midband) $=$ _____

    Z_{in} (midband) $=$ _____

    Z_{out} (midband) $=$ _____

6. Using the transient mode, generate V_{out} and note the distortion. As a quantitative measure of the distortion, determine the following: (Reminder: Frequency-based harmonic distortion is the subject of Chapter 23.)

    $$\% \text{ distortion} = \frac{\text{Vpeak(difference)}}{\text{Vpeak(average)}} \times 100 = \underline{\qquad}$$

Buffer

7. Draw the JFET buffer circuit of Figure 20.3 and set the attributes as shown.

8. Using the following equations, calculate by hand the "big three." (Because we are still using half-current biasing, g_m at the Q point is the same as before.)

    $$A = \frac{R_{B1}\|R_{B2}}{R_L\|R_S + 1/g_m} = \underline{\qquad}$$

    Z_{in} $=$ $R_{B1}\|R_{B2}\|Z_{in}$(FET) $=$ _____

    Z_{out} $=$ $R_S\|1/g_m$ $=$ _____

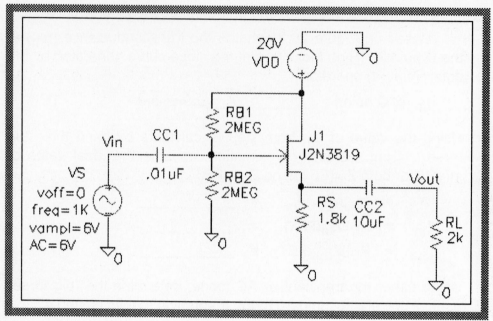

FIGURE 20.3

JFET buffer

9. Using PSpice (transient or AC mode), determine the "big three." Compare your answer to the theoretical results of step 8.

$$A \text{ (midband)} = \underline{\hspace{3cm}}$$
$$Z_{in} \text{ (midband)} = \underline{\hspace{3cm}}$$
$$Z_{out} \text{ (midband)} = \underline{\hspace{3cm}}$$

10. Based on the transient output waveform, determine the degree of distortion from the following equation:

$$\% \text{ distortion} = \frac{Vpeak(difference)}{Vpeak(average)} \times 100 = \underline{\hspace{2cm}}$$

Advanced Activities

11. Based on the results of Figure 20.2, develop an equation for transconductance (g_m) in terms of V_{GS}, V_{GSoff}, and a constant. (<u>Hint</u>: The equation is linear.)

EXERCISES

- Analyze the circuit of Figure 20.4 and predict the waveforms at the gate and drain of the FET. Generate the waveforms using PSpice and compare to your predictions. (<u>Hint</u>: The circuit is called a "chopper.")

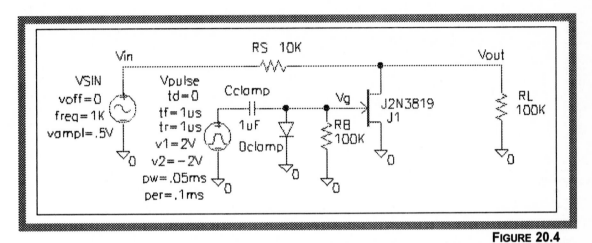

FIGURE 20.4

JFET chopper

QUESTIONS & PROBLEMS

1. Why is the voltage gain (A) of a FET amplifier generally less than that of a bipolar amplifier?

2. Referring to Figures 20.1 and 20.2, why would the voltage gain go up if the value of R_S is decreased?

3.   Looking to Figure 20.1, what allows resistor R_S to be so large?   What advantage is there to a large value of R_S?

4.   Why is the power gain of a FET circuit very large?

5.   Summarize the similarities and differences between bipolar and FET amplifiers and buffers.

PART V

Special Solid-State Studies

In Part V we examine several special applications of solid-state devices. We will find that the transistor can be used as a switch for digital applications, and that a special four-layer device can act as a latch.

PART IV

Special Solid-State Studies

CHAPTER 21

The Transistor as a Switch
Frequency of Operation

OBJECTIVES

- To design and analyze the bipolar transistor and FET when used as a switch.
- To test methods of increasing the frequency of operation.

DISCUSSION

Most of the previous chapters have concentrated on the transistor as used in analog (linear) applications. The other side of the coin is digital (nonlinear) applications. In a digital application, the transistor is a switch that operates between high and low states.

- When the bipolar transistor is used as a switch, the high and low states usually correspond to saturation and cutoff.

- When the FET is used as a switch, the high and low states usually correspond to I_{DSS} (maximum current) and V_{GSoff} (no current).

SWITCHING SPEED

An important consideration in digital circuits is the time it takes to switch between states. The faster the switching speed, the higher the frequency of operation. For the bipolar transistor, we define the terms of Figure 21.1 as follows:

- t_s (*storage time*) is the time required to come out of saturation (0% to 10%).

- t_r (*rise time)* is the time required to make the transition from saturation to cutoff (10% to 90%).

- t_d (*delay time*) is the time required to come out of cutoff (100% to 90%).

- t_f (*fall time*) is the time required to make the transition from cutoff to saturation (90% to 10%).

For the 3904 transistor, the spec sheet lists these terms as follows:

t_s = 200ns t_r = 35ns t_d = 35ns t_f = 50ns

The maximum frequency of operation [$1/(t_d + t_r + t_s + t_f)$] is therefore 3.125MHz.

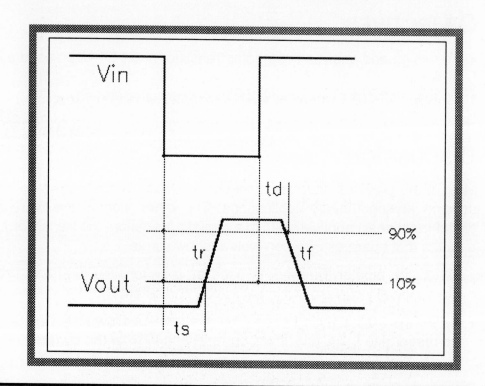

FIGURE 21.1

Switching time definitions

SIMULATION PRACTICE

The Bipolar Switch

1. Draw the basic transistor switch of Figure 21.2. (<u>Note</u>: The values of R_B and R_C were chosen to match the spec sheet test conditions of I_CMAX = 10mA and I_BMAX = 1mA.)

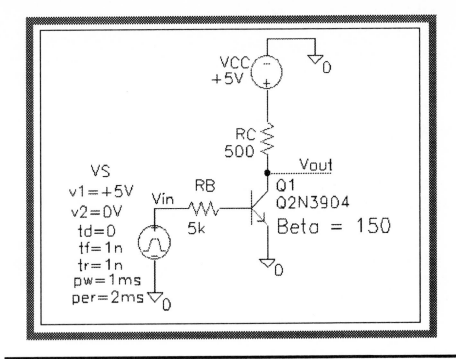

FIGURE 21.2

Basic bipolar
transistor switch

2.   Run PSpice and generate the waveforms of Figure 21.3. (Because
of a slight overshoot, you may have to adjust the Y-axes for 0 to
5V range.)

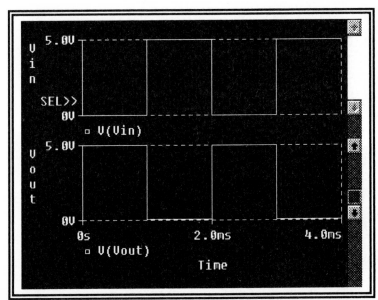

FIGURE 21.3

Bipolar switch
waveforms

3. Based on the waveforms of Figure 21.3, why is the switch also called an *inverter*?

4. Judging from the waveforms of Figure 21.3, does the transistor properly operate between saturation and cutoff?

    Yes   No

BJT Switching Time

5. Looking at the waveforms of Figures 21.3, it appears the output changes "instantly" with the input. However, increase the frequency by a factor of 1000 (pw = 1µs, per = 2µs), and generate the curves of Figure 21.4.

6. Using the cursor, determine values for each of the following and compare to the spec sheet values listed earlier in the discussion. (Note: Because our test conditions are not precisely the same as the spec sheet test conditions, do not expect close correlation between the PSpice results and the spec sheet results.)

    t_s = _____ t_d = _____ t_r = _____ t_f = _____

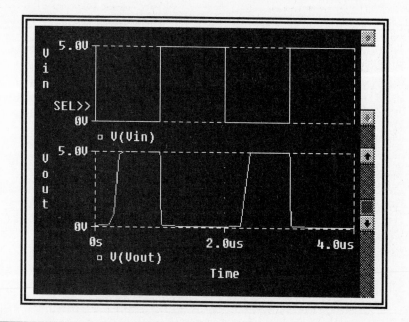

FIGURE 21.4

High-speed waveforms
showing delays

PSpice for Windows

7. To reduce the switching times [especially the storage (t_s) and delay (t_d) times], add the *speedup capacitor* of Figure 21.5. (The speedup capacitor bypasses resistor R_B during *changes*, thereby allowing electrons to move more quickly into and out of the base.)

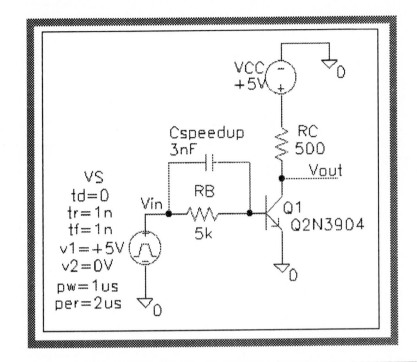

FIGURE 21.5

Adding a "speedup" capacitor

8. Generate new waveforms and note the storage, delay, and transition times. Were they greatly reduced?

   Yes   **No**

Advanced Activities

9. Using the equation developed in the discussion, analyze the waveforms generated by step 8 and determine the highest frequency of operation. By increasing the frequency in steps, verify your findings using PSpice.

10. By generating waveforms, compare the base current between the regular and "speedup" switches (Figures 21.2 and 21.5). Comment on the results.

PSpice for Windows

11. Draw the JFET switch of Figure 21.6, determine an appropriate value of R_D, and generate waveforms. (<u>Hint</u>: The transistor should be driven between cutoff and I_{DSS}.)

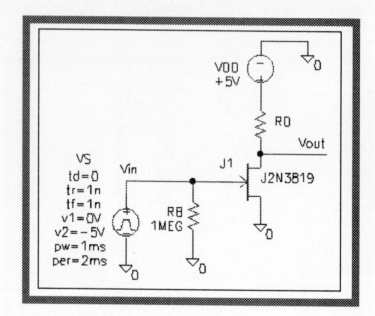

FIGURE 21.6

JFET switch

EXERCISES

- Using PSpice, determine the maximum frequency of operation of the CMOS inverter of Figure 21.7. (Could this basic inverter be part of a 66MEGHz 486 microprocessor?)

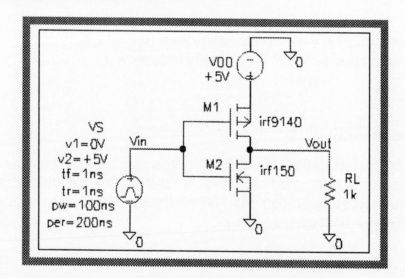

FIGURE 21.7

CMOS inverter circuit

- Determine the input/output characteristics and maximum frequency of operation of the TTL (*transistor-transistor logic*) inverter of Figure 21.8. Place speedup capacitors about R1, R2, and R4 and again determine the maximum frequency of operation.

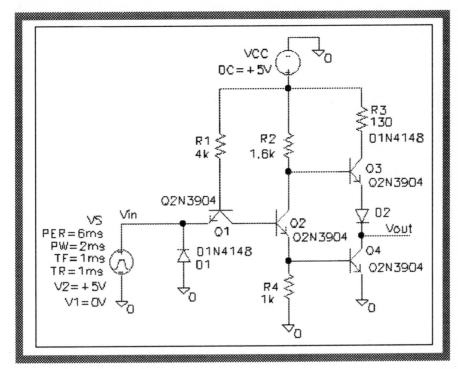

FIGURE 21.8

TTL inverter circuit

- The ECL (*emitter-coupled logic*) switch (inverter) shown in Figure 21.9 increases the switching speed by avoiding saturation. Explain how the circuit works. Using PSpice, determine its characteristics and maximum frequency of operation.

QUESTIONS AND PROBLEMS

1. When a transistor enters saturation, what happens to the following? (Enter "up" or "down" after each term.)

    (a) V_{CE} goes _____
    (b) *Beta goes* _____
    (c) I_C goes _____

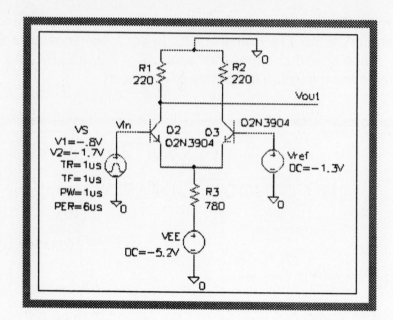

FIGURE 21.9

ECL inverter
circuit

2. Electrons that have "saturated" the base are swept out of the base during:

    t_d t_r t_s t_f

3. Why does the "speedup" capacitor of Figure 21.5 reduce the delay (t_d) and storage (t_s) times?

4. On the following load line graph, circle the area in which analog (linear) circuits normally operate, and the areas in which digital circuits normally operate (except when in transition).

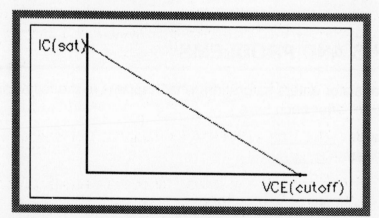

CHAPTER 22

Thyristors
The Silicon-control Rectifier

OBJECTIVES

- To determine the "on" and "off" characteristics of the SCR.
- To use the SCR in the design of an efficient power controller.

DISCUSSION

Thyristors are used for special switching applications. The most common thyristor is the *silicon-control rectifier* (SCR), a device that acts as a latch. When turned on, it tends to stay on; when turned off, it tends to stay off.

Turning to Figure 22.1(a), an SCR is a four-layer PNPN device. As shown by the equivalent circuit of Figure 22.1(b), it acts as overlapping NPN and PNP transistors. Because the collector of one transistor feeds the base of the other, both transistors must be on or both must be off. (It is impossible for one transistor to be on and the other off, except during the brief time when they are changing state.)

The schematic symbol for the SCR is given in Figure 22.1(c). (Because the SCR is made up of PSpice primitives, it is a *subcircuit* and therefore carries the "X" designator.)

We will use the test circuit of Figure 22.2 to demonstrate the SCR's characteristics.

- The SCR is turned on by a threshold combination of anode (A) voltage and gate (G) current.

- The SCR is turned off when the forward anode-to-cathode current drops below the *holding current* (I_H).

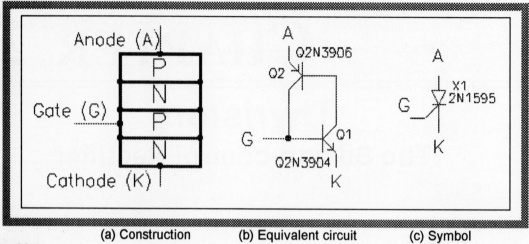

(a) Construction   (b) Equivalent circuit   (c) Symbol

FIGURE 22.1

The SCR
(a) Construction
(b) Equivalent circuit
(c) Symbol

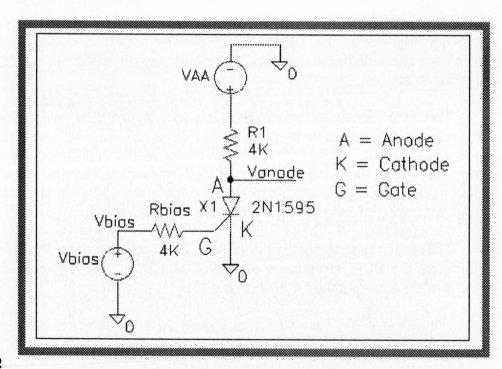

A = Anode
K = Cathode
G = Gate

FIGURE 22.2

SCR test circuit

SIMULATION PRACTICE

1. Draw the test circuit of Figure 22.2 and set the attributes as shown. (Because both V_{bias} and V_{AA} will be swept, DC bias point values need not be assigned at this time.)

SCR Operating Curves

2. Generate the SCR *operating curves* of Figure 22.3. (Refer to the *process summary* below.)

Process Summary for Generating SCR Operating Curves

- The <u>Main Sweep</u> variable is V_{AA}, generated by a DC Main Sweep from 0 to 55V in increments of 1V. The <u>Nested Sweep</u> variable is Vbias, generated by a DC Nested Sweep of value list 8.35, 8.5, and 8.75.

- The <u>X-axis</u> variable is SCR anode voltage [V(Vanode)], and the <u>Y-axis</u> is negative anode current [–I(R1)]. A second Y-axis is gate current [I(Rbias)].

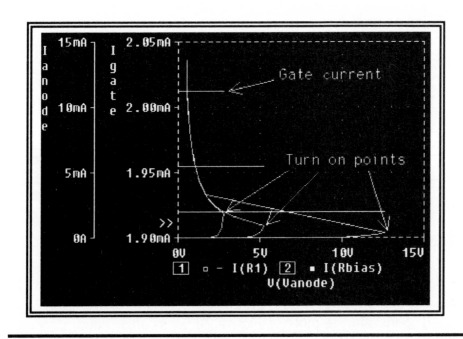

FIGURE 22.3

SCR operating curves

3. List below the three pairs of anode voltages [V(Vanode)] and gate currents [I(Rbias)] shown in Figure 22.3 that will turn on the SCR.

	V(Vanode)	I(Rbias)
Pair 1:		
Pair 2:		
Pair 3:		

Power Controller

4. Draw the circuit of Figure 22.4, which shows the SCR used as a *power controller*. (Resistor Rcontrol varies the percentage of time that the SCR is on.)

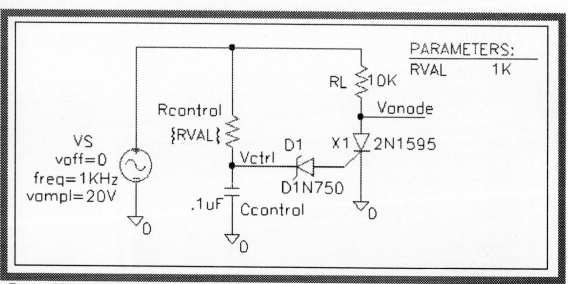

FIGURE 22.4

SCR power control

5. By sweeping Rcontrol as a parameter (values 100Ω, 2kΩ, and 5kΩ), generate the plots of Figure 22.5. In each case, circle the area when the SCR is on. (Note the use of the "@" operator.)

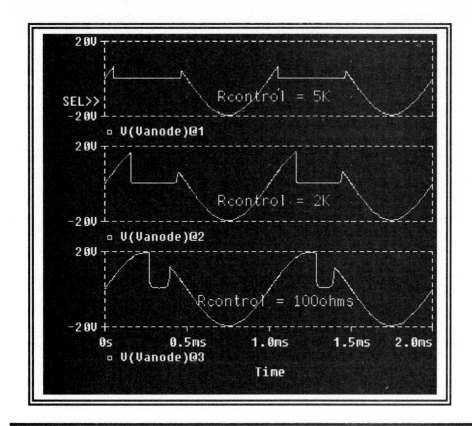

FIGURE 22.5

SCR power control waveforms

6. Based on Figure 22.5, what value of Rcontrol delivers the greatest power to the load? (Circle your answer.)

   100Ω   **2kΩ**   **5kΩ**

7. To the power control waveforms of Figure 22.5, add graphs of control voltage [V(Vctrl)@1, etc.] and note the trigger (turn on) points. Is the phase change between plots apparent?

   Yes   **No**

Advanced Activities

8. For the Rcontrol = 5kΩ case, determine the approximate holding current. (<u>Hint</u>: What is the current when the SCR *starts* to turn off?)

EXERCISES

• Substitute the SCR equivalent circuit of Figure 22.1(b) and re-generate the curves of Figure 22.5. Is the equivalent circuit similar in function to the SCR?

• Using the 2N5444 triac, redesign the power control circuit of Figure 22.4 to achieve full-wave control, and generate waveforms similar to Figure 22.5.

QUESTIONS & PROBLEMS

1. How do you turn an SCR on, and how do you turn it off?

2. Based on Figure 22.3, if the gate current is 1.94mA, approximately what anode voltage would turn on the SCR?

3. Referring to the SCR equivalent circuit of Figure 22.1(b), why are the transistors either both on or both off?

4. Why is the SCR an efficient method of controlling power (such as in Figure 22.4)?

5. Looking at the following two symbols, what is the difference between the
    SCR and the triac?

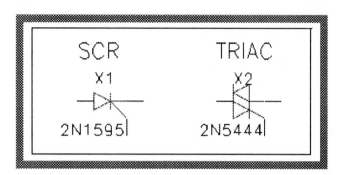

PART VI

Special Processes

In Part VI we introduce a number of special processes available under PSpice. These include *harmonic distortion*, *noise analysis*, *Monte Carlo analysis*, *worst case analysis*, *performance analysis*, *behavioral modeling*, and *hierarchy*.

We will find that PSpice calculates and presents in seconds results that would take many hours using hand analysis and calculation on actual circuits.

CHAPTER 23

Harmonic Distortion
Fourier Analysis

OBJECTIVES

- To show how nonlinear circuit elements lead to harmonic distortion.
- To analyze repetitive waveforms in the frequency domain.
- To determine total harmonic distortion in an amplifier.

DISCUSSION

As we have seen previously, transistors are inherently *nonlinear* devices. This is because their transconductance curves (Figure 23.1) are not straight lines. When used in amplifiers, this leads to *nonlinear distortion* (Figure 23.2).

Although the distortion in the time domain is clear, in many cases it is more convenient and instructive to view the distortion in the frequency domain (the *harmonic* distortion). The process of converting from the time domain to the frequency domain is known as *Fourier analysis*, and is the subject of this chapter.

For continuous signals, we use the *continuous Fourier transform* (CFT) shown here:

Frequency-to-time   **Time-to-frequency**

$$A(t) = \int_{-\infty}^{\infty} A(f)e^{j2\pi ft}df \qquad A(f) = \int_{-\infty}^{\infty} A(t)e^{-j2\pi ft}dt$$

When the waveform is sampled or analyzed on a digital computer, we must adopt the *discrete Fourier transform* (DFT):

$$a(k) = \sum_{j=0}^{N-1} A(j)e^{i2\pi jk/N} \qquad A(j) = \frac{1}{N}\sum_{k=0}^{N-1} a(k)e^{-i2\pi jk/N}$$

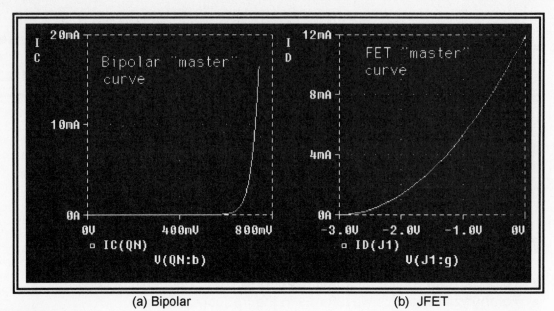

(a) Bipolar   (b) JFET

FIGURE 23.1

Transconductance
curves
(a) Bipolar
(b) JFET

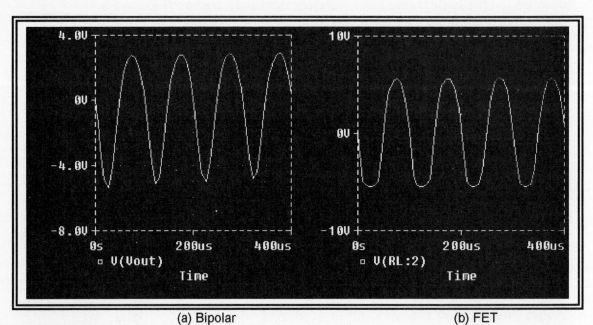

(a) Bipolar   (b) FET

FIGURE 23.2

Output waveforms
(a) Bipolar
(b) FET

If the time-domain waveform is repetitive, a great simplification results: The Fourier frequency components are harmonically related (multiples of the fundamental frequency). This means that the discrete Fourier transform can be performed by *fast Fourier transform* (FFT) techniques, resulting in a great reduction in calculation time.

Because all waveforms are assumed to be repetitive, PSpice offers the following two types of Fourier analyses:

- A DFT performed by PSpice on the last complete cycle of a specified voltage or current, with detailed harmonics tabulated in the output file.

- A FFT performed by Probe on any transient expression and displayed as a graph.

SIMULATION PRACTICE

1. Bring back the amplifier circuit shown in Figure 23.3 (first developed in Chapter 14).

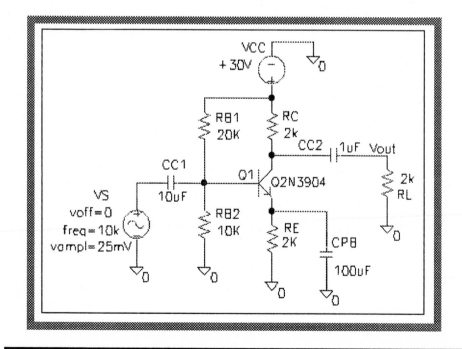

FIGURE 23.3

Bipolar amplifier circuit

PSpice for Windows

The Discrete Fourier Transform

<u>Reminder</u>: The results of a DFT are sent only to the output file, and are not sent to Probe for graphing.

2. Bring up the Transient dialog box and fill it in as shown in Figure 23.4. (We sweep from 0 to .4ms to provide four complete cycles of the 10kΩ input signal.) The Fourier Analysis section tells the system to perform a DFT on V_{out}, using a center frequency of 10kHz (the input frequency), and to generate four harmonics for tabulation in the output file.

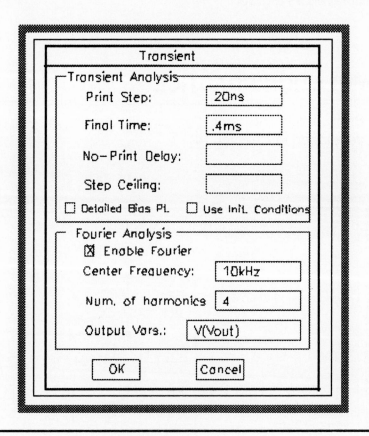

FIGURE 23.4

Transient dialog box

3. Run PSpice, bring up the output file (**Analysis**, **Examine Output**), and scroll to the Fourier data summarized by Table 23.1.

FOURIER COMPONENTS OF TRANSIENT RESPONSE V(Vout)
DC COMPONENT = −2.839359E−01

HARMONIC NO	FREQ (HZ)	FOURIER COMPONENT	NORMALIZED COMPONENT	PHASE (DEG)	NORMALIZED PHASE (DEG)
1	1.000E+4	3.761E+00	1.000E+00	−1.784E+02	0.000E+00
2	2.000E+4	6.613E−01	1.758E−01	9.451E+01	2.729E+02
3	3.000E+4	5.063E−02	1.346E−02	2.003E+01	1.984E+02
4	4.000E+4	9.298E−03	2.472E−03	1.086E+02	2.870E+02

TOTAL HARMONIC DISTORTION = 1.763515E+01 PERCENT

TABLE 23.1

Detailed frequency, amplitude, and phase of each harmonic

4. Based on the results of Table 23.1, which harmonic components carry significant energy?

   1   2   3   4

5. The *total harmonic distortion* shown in Table 23.1 is calculated by taking the square root of the sum of the squares of the harmonic components and reporting the result as a percentage of the fundamental component.

   The following equation uses values from Table 23.1. Using your calculator, verify the value determined by PSpice (17.64%).

$$\frac{(.6613^2 + .0506^2 + .0093^2)^{1/2}}{3.761} \times 100 = \underline{\hspace{2cm}}$$

The Fast Fourier Transform

Reminder: The results of a FFT appear only as a Probe graph, and are not sent to the output file.

6. Bring the default Probe graph window to the forefront, and convert the X-axis to frequency in order to display the Fourier components (**Plot**, **X-Axis Settings**, **Fourier**, **OK**).

7. Display the output voltage [V(Vout)], expand the X-axis, and generate the graph of Figure 23.5.

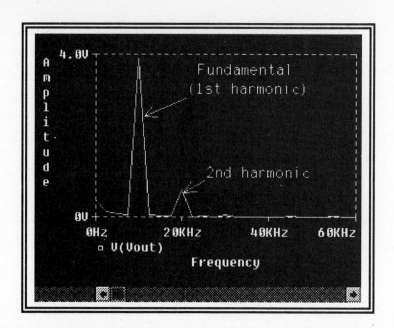

FIGURE 23.5

Fourier display of
the output voltage

8. Looking at the graph, *approximately* how many harmonics are visible? How many carry significant energy?

   Number of harmonics visible = _____

   Number carrying significant energy = _____

9. Does the FFT-generated graph of Figure 23.5 generally give the same results as the DFT-generated table of 23.1? (<u>Hint</u>: Measure the amplitude of the DC and harmonic components and compare to Table 23.1. Ignore the minus sign on the DFT DC component.)

   Yes   No

10. If you wish, perform a FFT on any other available voltage or current.

11. Add a 25Ω swamping resistor to the amplifier of Figure 23.3 (and increase V_{in} to 100mV).

   (a) Using the DFT method, what is the total harmonic distortion and by what percentage is it reduced from the unswamped case?

   Total harmonic distortion (swamped case) = _____

   Percent reduction = _____

(b) Using the FFT method, display the output's Fourier components and compare to the unswamped case. Are all harmonics (but the 1st) greatly reduced in amplitude?

Yes   **No**

Advanced Activities

12. Perform a complete harmonic analysis of the JFET amplifier of Figure 23.6 and compare to the bipolar results.

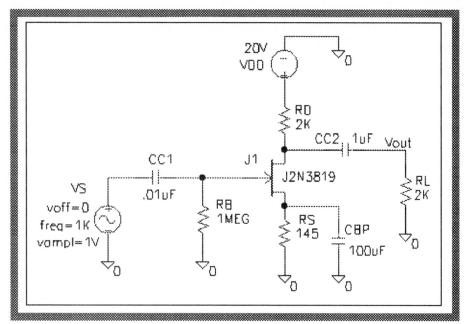

FIGURE 23.6

JFET amplifier

13. Using the data from Table 23.1, we generate the superposition equation shown below for the DC and first two harmonics. (Harmonic components 3 and 4 are not significant.)

> −.284 + 3.76*sin(6.28*10k*Time − 178.4*6.28/360)
> + .661*sin(6.28*20k*Time + 94.5*6.28/360)

Using PSpice, display the corresponding time-domain curve. (**Trace**, **Add**, enter the equation into the *Trace Command Box*.) Does the resulting trace resemble the original waveform of Figure 23.2(a)?

14. The resolution of the FFT depends on the number of cycles evaluated.

   To demonstrate this, restrict the FFT evaluation to the last two cycles (200µs to 400µs): **Plot**, **X-axis Settings**, **Restricted** (under *Use Data*), fill in 200us to 400us, **OK**. How does the output compare to the unrestricted case (0µs to 400µs)? (<u>Note</u>: Use either the original or swamped circuit.)

15. Besides the *number* of cycles that are sampled (step 14), a DFT and FFT also depend on *which* cycles are sampled.

   (a) Under Probe, shift the FFT evaluation time to 0µs to 200µs (from 200µs to 400µs) and note the change from step 14.

   (b) Under PSpice, change the transient Final Time from .4ms to .2ms. Compare the new Fourier table in the output file with Table 23.1. (Remember that the DFT uses only the last complete cycle in its transform calculations.)

EXERCISES

- Determine the total harmonic distortion of the audio amplifier of Figure 17.8. How would you improve (lower) the distortion value?

QUESTIONS & PROBLEMS

1. Why does a transistor cause harmonic distortion? Why does a resistor *not* cause harmonic distortion.

2. Under what conditions are the frequency components of a waveform harmonically related (multiples of the fundamental frequency)?

3. Based on the harmonic content of the first four harmonics (Table 23.1), were we justified in limiting the analysis to four harmonics?   Would be be justified in limiting the analysis to two harmonics?

4. Why does a buffer normally have a smaller total harmonic distortion than an amplifier?   (<u>Hint</u>: Does a buffer have "built-in" swamping?)

5. Referring to Figure 23.2, why is the DC component negative (as indicated by Table 23.1)?

6. Comparing Table 23.1 and Figure 23.5, does it appear that the FFT displays the DC component as an absolute number?

7. Looking at Table 23.1, how is the normalized value obtained from the component value?

8. From step 13, we have a good idea how a Fourier transform switches from the frequency domain to the time domain (component waveforms are superimposed). Explain how the opposite takes place. That is, how does a Fourier transform go from a time-domain curve to a frequency-domain curve? (<u>Hint</u>: Could we use test sine waves of different frequencies and look for net constructive or destructive interference.)

CHAPTER 24

Noise Analysis
Signal-to-Noise Ratio

OBJECTIVES

- To perform a noise analysis on an amplifier.
- To determine the *signal-to-noise ratio*.

DISCUSSION

The noise-generating devices in a circuit are the resistors and the semiconductor devices. A *noise* analysis tells the designer how the noise from all such devices will affect an output signal.

For example, from Figure 24.1, suppose we wish to determine the *signal-to-noise ratio* at the output node. A noise analysis will give us the answer.

Noise analysis can only be done in conjunction with an AC analysis. The noise function generates a noise "density" spectrum for each device over a range of frequencies and performs an RMS sum at the specified output node. Also reported is the "equivalent" noise from a specified input source that would cause the same output noise value if injected into a noiseless circuit.

SIMULATION PRACTICE

1. Draw (or bring back from Chapter 14) the amplifier circuit of Figure 24.1.

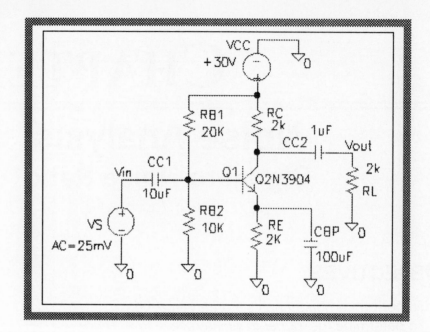

FIGURE 24.1

Amplifier test
circuit

2. Bring up the *AC Sweep and Noise Analysis* dialog box (**AC Sweep**), fill in as shown in Figure 24.2, **OK**, **Close**.

- *Output Voltage* [V(Vout)] gives the node at which the AC noise is to be determined.

- *I/V Source* (VS) is the independent voltage or current source at which the equivalent input noise will be calculated.

- *Interval* (10) causes a detailed table to be printed to the output file for every tenth frequency. (If no value is specified, no tables will be generated.)

3. Run the analysis (**Analysis**, **Simulate**) and generate the *noise density* (volts/root Hz) plots of Figure 24.3. (<u>Note</u>: Volts/root Hz = volts/Hz$^{1/2}$.)

Y-axis 1 shows ONOISE (output noise density)—the RMS summed noise (volts/root Hz) at the output node [V(Vout)]. Y-axis 2 shows INOISE (input noise density)—the equivalent RMS input noise (volts/root Hz) at V_{in} (V_S).

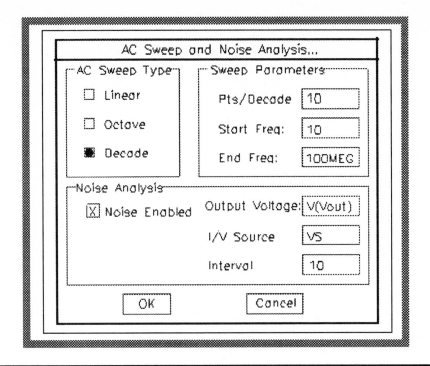

FIGURE 24.2

AC Sweep and
Noise Analysis
dialog box

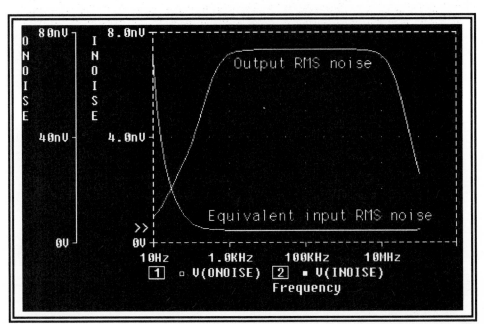

FIGURE 24.3

Noise plots

4. Based on the results (Figure 24.3), what is the output noise density at 100kHz? At this same frequency, what is the equivalent noise density at the input (V_S)?

> **VRMS/Hz(Onoise)@100kHz = _____**
>
> **VRMS/Hz(Inoise)@100kHz = _____**

5. As shown in Figure 24.4, add a second plot and display the input and output *signal* voltages.

> What is the *signal-to-noise* ratio at the output? For example, at 100kHz, what is the ratio of output signal voltage to output noise voltage expressed in both regular and dB format?
>
> **Signal-to-noise ratio @100kHz (regular) = _____**
>
> **Signal-to-noise ratio @100kHz (dB) = _____**

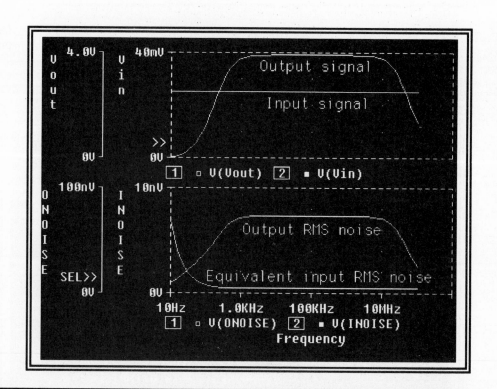

FIGURE 24.4

Comparing signal-to-noise ratios

6. Examine the *noise analysis* section of the output file. Per our instructions (Figure 24.2), was a separate noise analysis tabulated in detail for each of the frequencies 10Hz, 100Hz, 1kHz, 10kHz, 100kHz, 1MEGHz and 10MEGHz?

    Yes   No

7. Table 24.1 shows the noise analysis section in the output file for 100kHz.

    (a) Does most of the noise come from the transistor or from the resistors? (Note: RB, RC, etc., are transistor parameters; R_RC, R_RB1, etc., are resistors.)

    (b) Which resistor or resistors contribute the greatest noise?

    (c) Is the *equivalent input noise at VS* (4.699E-10 V/RT HZ) times the transfer function gain (156.3) equal to the *total output noise voltage* (7.346E-8 V/RT HZ)?

         Yes   No

```
****  NOISE ANALYSIS        TEMPERATURE =  27.000 DEG C
****************************************************************
                FREQUENCY =  1.000E+05 HZ
     **** TRANSISTOR SQUARED NOISE VOLTAGES (SQ V/HZ)
                        Q_Q1
                RB      4.051E-15
                RC      4.259E-23
                RE      0.000E+00
                IB      2.014E-17
                IC      1.310E-15
                FN      0.000E+00
              TOTAL     5.381E-15

     **** RESISTOR SQUARED NOISE VOLTAGES (SQ V/HZ)
        R_RC     R_RB1     R_RL      R_RB2     R_RE
TOTAL  7.470E-18 5.130E-22 7.470E-18 1.026E-21 5.134E-23
**** TOTAL OUTPUT NOISE VOLTAGE     = 5.396E-15 SQ V/HZ
                                    = 7.346E-08 V/RT HZ
            TRANSFER FUNCTION VALUE:
         V(Vout)/V_VS           = 1.563E+02
    EQUIVALENT INPUT NOISE AT V_VS = 4.699E-10 V/RT HZ
```

TABLE 24.1

Noise analysis section of output file at 100kHz

Advanced Activities

8. Increase the temperature to 100° (212°F) and again determine the total noise at 100kHz. (How does it compare to the noise at the default temperature of 27°?)

> Total V/RT HZ (noise) @ 100kHz and 27° = _____

> Total V/RT HZ (noise) @ 100kHz and 100° = _____

EXERCISES

-   Perform a noise analysis on the audio amplifier of Figure 17.8. At 100kHz, how does the signal-to-noise ratio (dB) compare with that of the single-stage amplifier of this chapter?

QUESTIONS & PROBLEMS

1. Perform a root-mean-square (RMS) of the numbers below:

> RMS of 7, 2, -4, 8, 12 = _____

2. What causes noise in a resistor?

3. Based on Figure 24.3, what is the "noise bandwidth" of the amplifier circuit of Figure 24.1?

4. Looking to Figure 24.3, why does the "equivalent input RMS noise" rise significantly at lower frequencies?   (Hint: What roll do the capacitors play?)

5.   Using the ratio of resistors method (rc||rl/re', where re' = 25mV/IEQ), calculate the voltage gain of the amplifier of this chapter. How does this compare to the *TRANSFER FUNCTION VALUE* of Table 24.1 (156.3)?

6.   If it turns out that the signal-to-noise ratio of the amplifier of Figure 24.1 is too low (too much noise), what would be the most logical step to increase the ratio?   (<u>Hint</u>: Looking at Table 24.1, does the transistor give the greatest noise component?)

7.   Referring to Table 24.1, how is 7.346E–08 V/RT HZ obtained from 5.396E–15 SQ V/HZ?

CHAPTER 25

Monte Carlo Analysis
Tolerances

OBJECTIVES

- To assign tolerance values to components.
- To perform a Monte Carlo analysis on an amplifier.

DISCUSSION

All the components in the amplifier of Figure 25.1 are assumed to be constants. For example, all the resistors have *exactly* the values indicated by their color code—and they never change. In the real world however, all resistors have *tolerances*, which specify how they might vary from their nominal value.

To determine the effects of such tolerance variations, PSpice offers *Monte Carlo* analysis.

MONTE CARLO

During a Monte Carlo analysis, PSpice performs several "runs" of a DC, AC, or transient analysis, each time varying model parameter values in a random manner within the tolerance range. It is this random nature that gives the process its "Monte Carlo" tag. The first run is always the "nominal" run, using the component's "face" value, with no tolerance variations.

Output is sent to *Probe* for graphical display and to the *output file* for tabular display.

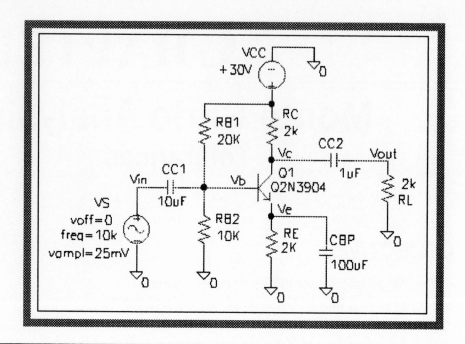

FIGURE 25.1

Small-signal
amplifier

MODELS AND TOLERANCES

Consider the simplest device in the amplifier of Figure 25.1—the *resistor* (part R). Because part R does not provide a library of model parameters, it cannot be assigned a tolerance parameter. However, part *Rbreak*, found in model library *breakout.lib*, does allow the addition of model parameters.

To part *Rbreak* we can assign *lot* and *device* tolerance parameters. A device tolerance causes devices that use the same model to vary *independently* of one another, and is appropriate for discrete devices. A lot tolerance causes all devices that use the same model to vary together, and are appropriate for integrated circuit devices. Any device can be assigned a combination of device and lot tolerances.

In this chapter, we will first perform a Monte Carlo analysis with only the resistors given tolerances. We will then add a tolerance value to a transistor parameter.

SIMULATION PRACTICE

1.   Draw the circuit of Figure 25.1 (from Chapter 14) and set all attributes as shown.

Define a Resistor Model

2.   Because part R does not support a model, we replace each *R* part with *Rbreak* as follows:

Hold the shift key down and **CLICKL** on all five resistors (to select all resistors at once), **Edit**, **Replace** (to bring up the *Replace Part* dialog box), enter Rbreak in *Replacement* box, if necessary **CLICKL** on *Keep Attribute Values* and *Selected Parts Only*, **OK**.

All five resistors now have the part and model name *Rbreak*, *with the model name displayed.*

> Because breakout parts are meant to have their model changed, their model name is visible on the schematic instead of their part name.

3.   The next step is to change the model name (a good idea in case the model is later made global), assign device and lot tolerances, and store the new modified model parameters to a local library.

To accomplish this, select any one of the five resistors, **Edit**, **Model**, **Edit Instance Model** to bring up the *Model Editor* dialog box, fill in as shown below (where R =1 is the scaling factor and Rmc is the new model name), **OK**.

. model Rmc RES
R=1 DEV=10% LOT=2%

The parameters for model *Rmc* are now stored in a local library (*.lib*) with the same file name as the schematic name.

4.   To reassign the remaining four resistors (presently model Rbreak) to the new model (Rmc), select the remaining four resistors, **Edit**, **Model**, **Change Model Reference**, change Rbreak to Rmc, **OK**, and *repeat the last three steps three more times*.

Your circuit should now resemble Figure 25.2.

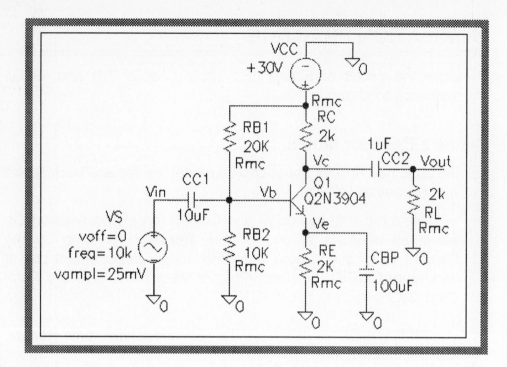

FIGURE 25.2

Circuit with resistor
tolerances set

> <u>Although we will not do so at this time</u>, we have the option of making the local model library global so all custom parts and models will be available to any schematic. This is accomplished as follows: **Analysis**, **Library and Include Files**, **Add Library***, enter library name (using a new library name is a good idea), **OK**.

Perform the Monte Carlo Analysis

5.   Set up the system for transient analysis from 0 to .2ms.

6.   To add Monte Carlo analysis on top of transient analysis: **Analysis**, **Setup**, **CLICKL** on *Monte Carlo/Worst Case* enabled box, **Monte Carlo/Worst Case** to bring up the *Monte Carlo or Worst Case* dialog box (Figure 25.3), and fill in as shown, **OK**, **Close**.

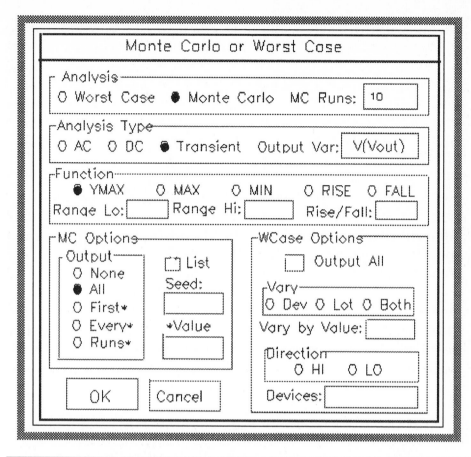

FIGURE 25.3

Monte Carlo
dialog box

- <u>Output var</u>: V(Vout) specifies the node at which the waveforms will be generated.

- <u>Function</u>: Specifies the function to be performed on the output variable waveform for each run. The results affect both the output file and the graphical presentation under *Probe.*

    YMAX finds the greatest difference from the nominal run, MAX and MIN finds the maximum and minimum value, RISE and FALL finds the first occurrence of the waveform crossing above or below the threshold value in the *rise/fall* field, and the *Range Lo:* and *Range HI:* boxes restrict the range over which a function will be evaluated. Only one function may be selected at a time.

- <u>MC Options</u> controls the runs. *None* causes only the nominal run to be produced, *All* causes all runs; *First* causes only the first *n* runs (*n* is placed in the *value* box), *Every* causes only every *n*th run, *Runs* causes only the runs listed in the "Value:" box, *List* prints to the output file the model parameter values used for each component during that run, and *Seed* defines the seed value for the random number generator (default is 17533).

7. Run the Transient/Monte Carlo analysis (**Analysis**, **Simulate**) and note the 10 runs performed in the PSpice run-time display box. After **All** (Available Sections), **OK**, the Probe graph appears with the default X-axis assigned the time range specified in the transient analysis setup.

8. Add the output variable and generate the graph of Figure 25.4. Note the 10 different waveforms that result from the statistical variation of all the resistors. (The first waveform is the nominal waveform in which all resistors are at their "face" value.)

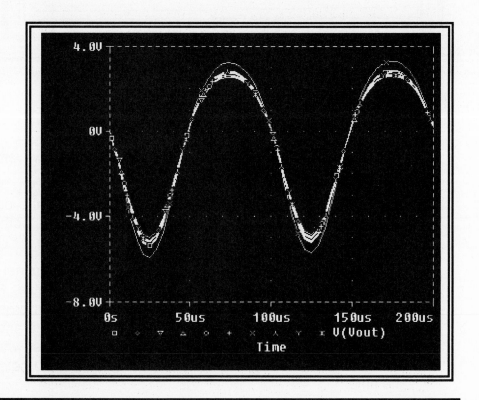

FIGURE 25.4

Monte Carlo
output waveform

PSpice for Windows

9. Looking at the results (Figure 25.4), what is the *approximate* percent variation between the lowest and highest peak values compared to the average peak value (@ 75µs)?

$$\% \text{ variation } = \frac{\textbf{Vpeak(difference)}}{\textbf{Vpeak(average)}} \times 100 = \underline{\hspace{2cm}}$$

The Output File

10. During a Monte Carlo analysis, a great deal of data is written to the output file. Examine the output file and find the INITIAL TRANSIENT SOLUTION for the *nominal* and nine additional runs. For example, Table 25.1 shows the initial transient data for the nominal and arbitrary run (pass) seven.

> Reminder: During the initial transient (bias point) solution at TIME = 0, all capacitors are opened and all inductors are shorted.

INITIAL TRANSIENT SOLUTION   TEMPERATURE = 27.000 DEG C
MONTE CARLO NOMINAL
****

NODE	VOLTAGE	NODE	VOLTAGE	NODE	VOLTAGE	NODE	VOLTAGE
(Vb)	9.8286	(Vc)	20.9250	(Ve)	9.1260	(Vcc)	30.0000
(Vin)	0.0000	(Vout)	0.0000				

VOLTAGE SOURCE CURRENTS
NAME   CURRENT
V_VCC   -5.546E-03
V_VS   0.000E+00
TOTAL POWER DISSIPATION 1.66E-01 WATTS

INITIAL TRANSIENT SOLUTION   TEMPERATURE = 27.000 DEG C
MONTE CARLO PASS 7
****

NODE	VOLTAGE	NODE	VOLTAGE	NODE	VOLTAGE	NODE	VOLTAGE
(Vb)	10.9990	(Vc)	18.7870	(Ve)	10.2900	(Vcc)	30.0000
(Vin)	0.0000	(Vout)	0.0000				

VOLTAGE SOURCE CURRENTS
NAME   CURRENT
V_VCC   -6.651E-03
V_VS   0.000E+00
TOTAL POWER DISSIPATION 2.00E-01 WATTS

TABLE 25.1

Initial transient solutions for nominal and pass 7.

11. Looking to Table 25.1, pass 7 produced what percent change in the initial transient collector voltage from nominal?

    % change from nominal = _____

12. Next, find the "sorted deviations" section of the output file (Table 25.2). In this table, all data is based on transient runs from TIME = 0 to 200μs with all capacitors and inductors taking on to their assigned values.

    The data listed for each run is in response to the **YMAX** function. For each run *past the nominal* (2 through 10), the maximum amplitude deviation from the nominal waveform is found and listed in order of magnitude as an absolute number, a percent of nominal, and a standard deviation (sigma). Also listed is the time that the maximum deviation occurred and whether the deviation was lower or higher than nominal.

    At the top of the table is the *overall* mean deviation and sigma. These values are based on all values in every sweep, and not just the single YMAX points.

```
SORTED DEVIATIONS OF V(Vout)   TEMPERATURE =  27.000 DEG C
                    MONTE CARLO SUMMARY
*********************************************************************

                       Mean Deviation  =  .1006
                            Sigma       =  .2467

     RUN                   MAX DEVIATION FROM NOMINAL
   Pass   7           .5565 (2.26 sigma) lower  at T =  25.1210E-06
                          ( 110.37% of Nominal)
   Pass   3           .3506 (1.42 sigma) higher at T =  25.1210E-06
                          ( 93.464% of Nominal)
   Pass   4           .2334 ( .95 sigma) higher at T =  25.1210E-06
                          ( 95.649% of Nominal)
   Pass   5           .2092 ( .85 sigma) higher at T =  25.1210E-06
                          ( 96.101% of Nominal)
   Pass  10           .2025 ( .82 sigma) higher at T =  25.1210E-06
                          ( 96.226% of Nominal)
   Pass   9           .1837 ( .74 sigma) higher at T =  25.1210E-06
                          ( 96.575% of Nominal)
   Pass   2           .1593 ( .65 sigma) higher at T = 125.1200E-06
                          ( 96.922% of Nominal)
   Pass   8           .0647 ( .26 sigma) higher at T = 173.1200E-06
                          ( 102.29% of Nominal)
   Pass   6           .0582 ( .24 sigma) higher at T =  21.1200E-06
                          ( 98.886% of Nominal)
```

TABLE 25.2

YMAX function
results

PSpice for Windows

13. Based on Table 25.2, what was the greatest deviation from nominal and when did it occur?

    Greatest deviation from nominal (%) = _____

    Found at time = _____ during run _____

Advanced Activities

14. Assign transistor parameter Bf (maximum forward ideal *Beta*) a 50% tolerance as follows: select Q1, **Edit**, **Model**, **Edit Instance Model**, add *DEV=50%* after *Bf=416.4*, **OK**. (<u>Note</u>: The transistor model is stored in a local library with same file name as the schematic and with extension *.lib*.)

    Generate a new waveform set and a new *Sorted Deviation* table, and compare to the previous results.

15. Plot V_C (the collector voltage) under Probe and compare the nominal and pass 7 values at TIME = 0 (the initial transient time) to Table 25.1.   Do they match?

EXERCISE

• Referring to Figure 25.1, change all resistors from low-cost (10% tolerance) to precision (1% tolerance) and perform another Monte Carlo analysis.   How do the two cases compare?

QUESTIONS & PROBLEMS

1. What is the difference between *device* and *lot* tolerances?

2. What is the difference between part names *R* and *Rbreak*?

3. Which *function* would you use to determine the first occurrence of the waveform crossing above the threshold value for each run?

4. How would you repeat a Monte Carlo analysis using a different sequence of random parameter values?

5. When doing a Monte Carlo analysis, the first run is always the _____ run.

6. Like the R component, the C (capacitor) component also does not have a model.   Based on the resistor case, how do you think a tolerance would be assigned to a capacitor?

7. What is the difference between a *local* and a *global* library?

8. The symbol for a given part is found in a library with extension _____, and the model for the part is found in a library with extension _____.

9. Define each of the following terms:

    (a) Initial transient.

    (b) Nominal.

CHAPTER 26

Worst Case Analysis
AC Sensitivity

OBJECTIVES

- To perform a sensitivity analysis on an amplifier.
- To perform a worst case analysis on an amplifier.

DISCUSSION

Although both Monte Carlo and worst case analysis make use of component tolerances and involve a number of "runs," they are quite different. Monte Carlo analysis, covered in the last chapter, varies component values in a random manner as the runs are made; Worst Case analysis does not use random variations at all.

Instead, worst case analysis involves a two-step process. First, we perform a *sensitivity* analysis, in which model parameters are varied one at a time for each device, with a DC, AC, or transient analysis run for each variation.

When the sensitivity analysis is done, PSpice uses the data to perform one final worst case run, with each parameter set up or down by its *full* tolerance in such a way as to produce the greatest output signal or the greatest deviation from nominal (or other result, depending on the function chosen).

SIMULATION PRACTICE

1. Bring back the amplifier in Figure 26.1 (reproduced from Chapter 25). (As set in Chapter 25, all resistors have a device and lot tolerance of 10% and 2%, and Q1 has no tolerance value set.)

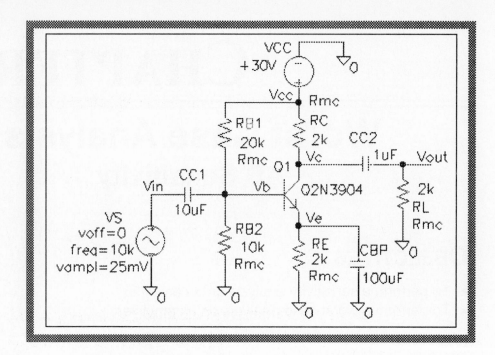

FIGURE 26.1

Amplifier ready for
worst case analysis

2. Bring back the *Monte Carlo or Worst Case* dialog box of Figure 26.2 (**Analyses**, **Setup**, **Monte Carlo/Worst Case**).

3. Switch to a worst case analysis (**Worst Case**), set *Analysis Type* to **Transient**, and set *Function* to **MAX** (to generate the maximum amplitude worst case output waveform) .

4. Fill in the *WCase Options* box as follows:

   • **Output All** to send data from all sensitivity runs to the output file. If not enabled, then only the nominal and final (worst-case) run generates output.

   • **Dev** to activate only the device tolerances.

   > For a more accurate worst case analysis, we should first perform an analysis with **Lot** selected, manually adjust the resistor values as specified, then perform another analysis with **Dev** selected. If **Both** is selected, all resistors are set up or down by the *same* amount, and the result is difficult to interpret.

   • *Vary by Value* is left blank because we wish to use the default method of varying parameter tolerances.

- **Hi** to specify which direction the worst case run is to go (relative to the nominal). (If *function* is YMAX or MAX, the default is HI; otherwise the default is LO.)

- Leave the *Devices* box blank so all devices will be included in the analysis.

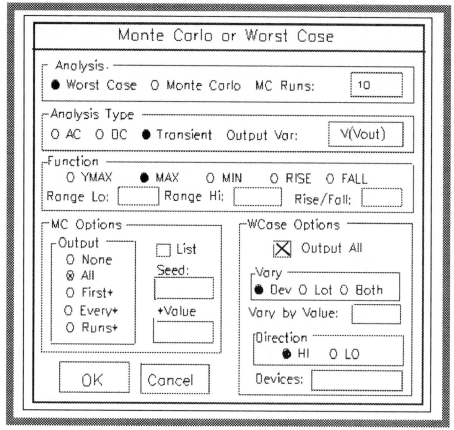

FIGURE 26.2

Monte Carlo or Worst
Case dialog box

5. Exit the *Monte Carlo* or *Worst Case* dialog box and return to the *Analysis Setup* menu (**OK**).

6. Make sure that the *Transient* analysis is selected, and that the final time is .2ms. Return to Schematics (**Close**).

7. Run the *Transient/Worst Case* analysis and wait for the *Available Sections* box to appear. **All** to include all devices and generate the initial graph, **OK**.

(<u>Note</u>: ...MINAL means the *nominal* waveform, ...L DEVICES means the waveform for *all devices*, and the other options specify waveforms to be generated for individual devices.)

8. Add the output trace [V(Vout)] to generate the individual parameter and final (worst case) output waveforms of Figure 26.3.

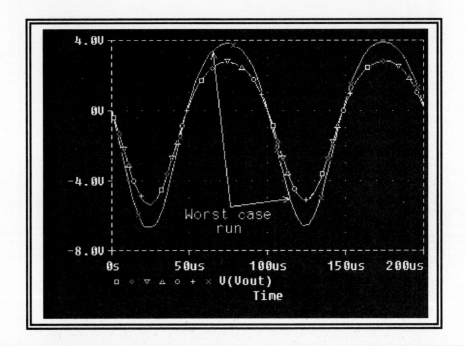

FIGURE 26.3

Worst case
waveform set

9. Looking at Figure 26.3, note that seven waveforms were drawn. In order of generation and listing, they are as follows:

    1 nominal waveform
    5 sensitivity waveforms for each resistor
    1 final worst case waveform

10. If you wish, zoom in on the six closely clustered nominal and sensitivity waveforms and verify that there are actually six separate waveforms.

11. Due to the MAX function, the worst case waveform (legend symbol ×) includes the maximum possible amplitude. Does the waveform show any evidence of clipping?

        Yes   **No**

Output File

12. Scan through the output file and locate the INITIAL TRANSIENT solution, generated for the nominal case and each of the five resistors. This represents the starting point (TIME = 0) for each of the six sensitivity runs.

13. Scan further into the output file and Locate the SENSITIVITY SUMMARY of Table 26.1, created using data from the six transient sensitivity runs—each from TIME = 0 to TIME = 200μs.

    As directed by the MAX function, each of the five resistors is increased in value one at a time for each run, and the maximum corresponding output voltage is located. The data is then compared to the nominal run, and the five resistors are ranked from most positive to most negative according to their sensitivity (their affect on the magnitude of the maximum output signal).

SORTED DEVIATIONS OF V(Vout)   TEMPERATURE = 27.000 DEG C
SENSITIVITY SUMMARY
****

RUN	MAXIMUM VALUE
R_RC Rmc R	2.8289 at T = 173.1200E-06
	(.4718% change per 1% change in Model Parameter)
R_RB2 Rmc R	2.8289 at T = 173.1200E-06
	(.4584% change per 1% change in Model Parameter)
R_RL Rmc R	2.8283 at T = 173.1200E-06
	(.2557% change per 1% change in Model Parameter)
NOMINAL	2.8276 at T = 173.1200E-06
R_RB1 Rmc R	2.8256 at T = 173.1200E-06
	(-.6898% change per 1% change in Model Parameter)
R_RE Rmc R	2.8249 at T = 173.1200E-06
	(-.9443% change per 1% change in Model Parameter)

TABLE 26.1

Sensitivity summary

14. Based on Table 26.1:

    (a) Which three resistors should be *increased* in order to increase the maximum output signal?

    _____   _____   _____

    (b) Which two resistors should be *decreased* in order to increase the maximum output signal?

    _____   _____

    (c) If the load resistor (R_RL) were increased by 10%, by what percentage would the maximum output voltage increase?

    % increase = _____

    (d) If you desired to reduce the chance of clipping at minimal cost by changing just one resistor's tolerance from 10% to 1%, which resistor would you choose? (Circle your answer.)

    RC   RB1   RL   RB2   RE

    Why did you choose this resistor?

15. Continue to scan through the output file and locate the UPDATED MODEL PARAMETERS data of Table 26.2—which shows how all resistor values are increased or decreased by their full tolerance amount (10%) during the final worst case run. (1.1 and .9 are scaling factors for the nominal values.)

    Does the data support your answers to step 14 (a) and (b)?

    Yes   No

UPDATED MODEL PARAMETERS		TEMPERATURE = 27.000 DEG	
WORST CASE ALL DEVICES			
DEVICE	**MODEL**	**PARAMETER**	**NEW VALUE**
R_RC	Rmc	R	1.1 (Increased)
R_RB1	Rmc	R	.9 (Decreased)
R_RL	Rmc	R	1.1 (Increased)
R_RB2	Rmc	R	1.1 (Increased)
R_RE	Rmc	R	.9 (Decreased)

TABLE 26.2

Worst case
parameter changes

16. When all tolerances are set as listed in Table 26.2, the system performs the final worst case run, and the summary data of Table 26.3 are written to the output file. As shown, the *maximum* output voltage was 3.8929V at 174.36µs, or 137.68% of the maximum nominal output voltage.

Do the results of Table 26.3 match the Probe data (the last run) of Figure 26.3?

Yes   No

```
┌─────────────────────────────────────────────────────────────┐
│                                                             │
│   SORTED DEVIATIONS OF V(Vout)   TEMPERATURE =  27.000 DEG C │
│                  WORST  CASE SUMMARY                         │
│  ***********************************************************  │
│              RUN              MAXIMUM VALUE                  │
│          ALL DEVICES      3.8929 at T =  174.3600E-06        │
│                           ( 137.68% of Nominal)             │
│                                                             │
│          NOMINAL          2.8276 at T =  173.1200E-06        │
│                                                             │
└─────────────────────────────────────────────────────────────┘
```

TABLE 26.3

Worst case summary

Advanced Activities

17. Assign transistor parameter *Bf* a 50% DEV tolerance and compare the outcome to the previous results (with no transistor tolerance value).

18. Assign the three capacitors of Figure 26.1 10% tolerance values (Cbreak) and compare the outcome to the previous results.

19. Repeat the worst case analysis using the YMAX function (instead of the MAX function). Explain the differences. (<u>Hint</u>: YMAX finds the greatest *difference* from nominal; MAX finds the greatest amplitude, without regard to nominal.)

EXERCISES

- Perform a *complete* worst case analysis on the audio amplifier of Figure 17.8. Set tolerances for the resistors, and include capacitor and transistor tolerances if you wish. Vary the function from YMAX to FALL. (Compare the results when using 10% and 1% resistors and capacitors.)

QUESTIONS & PROBLEMS

1. Which of the following analyses generate random numbers?
    (a) Monte Carlo
    (b) Sensitivity
    (c) Worst case

2. When performing a worst case analysis, why must a sensitivity analysis be done first?

3. Based on the sensitivity data of Table 26.1, the output voltage is *least* sensitive to which resistor?

4. Based on Table 26.1, by what percent would V_{out} change if RC changed by 20%?

5. Based on step 11, why is there a potential problem with the amplifier design of Figure 26.1?   If so, what could you do to solve the problem?

6. Pick any two resistors from Table 26.2 and explain why their increased or decreased values resulted in a larger maximum output voltage.

CHAPTER 27

Performance Analysis
Histograms

OBJECTIVES

- To use *performance analysis* to graph special functions.
- To generate a *histogram* from the results of a Monte Carlo analysis.

DISCUSSION

Performance analysis starts with a parametric analysis—multiple runs of a waveform while varying a parameter. For example, consider the tank circuit of Figure 27.1(a) (from Chapter 5) and the resulting parametric family of curves of Figure 27.1(b).

Suppose our task is to use this tank circuit in the design of an intermediate-frequency amplifier for an AM radio station. Let's assume that it must have a resonant frequency of 50kHz, a bandwidth of 8kHz, and a peak impedance (at resonance) between 1kΩ and 3kΩ.

Looking at Figure 27.1(b), we find that the circuit has the correct resonant frequency. The question is, *what value of Rtank results in a bandwidth of 8kHz and a peak impedance between 1kΩ and 3kΩ?*

We could estimate both answers from Figure 27.1(b)—but *performance analysis* provides a better way. With performance analysis, we can directly generate graphs of bandwidth and peak amplitude versus Rtank. Our radio station design solutions will be picked right off the graphs.

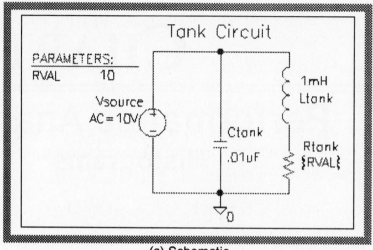

(a) Schematic

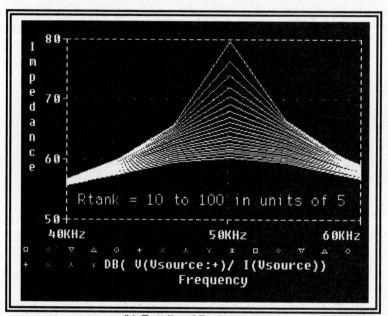

(b) Family of Bode curves

FIGURE 27.1

The tank circuit
(a) Schematic
(b) Family of Bode curves

THE GOAL FUNCTION

At the heart of performance analysis is the *goal function*. It is the goal function that analyzes each curve of a family of curves and generates a single value for each curve—which is then plotted against the parameter value.

We will use the two goal functions shown in Figure 27.2 to help us design our AM radio station.

- *Bandwidth* will search each successive impedance curve (representing each successive value of Rtank) and return the corresponding bandwidth for storage.

- *Amplitude* will search each successive impedance curve (representing each successive value of Rtank) and return the corresponding maximum Y-axis value for storage.

The stored values of bandwidth and impedance will then be plotted against Rtank.

```
bandwidth(1) = x2 - x1
{
      1|
            search LEVEL(70.7%,p) !1
            search LEVEL(70.7%,n) !2
            ;
}
```

```
amplitude(1) = y1
{
      1|
            search forward MAX!1
            ;
}
```

FIGURE 27.2

Goal functions
bandwidth
and *amplitude*

The major sections of the goal functions are as follows:

- *Bandwidth* and *amplitude* are the goal function names.

- 1| refers to the first *expression* (all the search commands between the 1| and semicolon). A goal function can have any number of expressions.

- = *x2 - x1* and = *y1* specify the returned values (to be stored and plotted against Rtank).

- *search LEVEL(70.7%,p)!1*, *search LEVEL(70.7%,n)!2*, and *search forward MAX !1* are *marked point expressions*. LEVEL searches for the positive-going (p) and negative-going (n) points that are 70.7% of the peak. MAX searches in the forward direction (left to right), looking for the maximum value.

When each searched-for point is found, for each corresponding value of Rtank, it is "marked" (stored) as coordinates x1 and y1. In the first case, X2 - X1 (the bandwidth) is returned, and in the second case, Y1 (the amplitude of the peak) is returned. When we plot the returned values against parameter variable Rtank, we pick the correct value of Rtank (corresponding to a bandwidth of 8kHz) right off the graph.

HISTOGRAMS

Once the correct value of Rtank is found, the next logical step is to assign Rtank a tolerance and note the effect on bandwidth. To accomplish this, we generate a *histogram*—a performance analysis done by varying Rtank randomly about its correct value according to the multiple runs of a Monte Carlo analysis.

With a histogram, the X-axis takes on the returned values of the goal function (bandwidth) and the Y-axis takes on units of percent. Using histograms, we will determine the percentage of time our tank circuit spends in various X-axis (bandwidth) "slots" due to the random tolerance variations of Rtank.

SIMULATION PRACTICE

1. Using a *text* editor, generate goal functions *bandwidth* and *amplitude* (both defined in the discussion). Store in file *probe.gf* in directory *msimev60* (the same directory as *probe.exe*).

   (*Caution*: Be sure to use an ASCII *text* editor, or a word processor in the text mode.)

2. Draw (or obtain from Chapter 5) the tank circuit of Figure 27.1(a).

3. Be sure the main sweep is an AC Sweep from 10kHz to 100kHz (30 pts/dec), and the nested (parametric) sweep is Rtank, having global values (RVAL) from 10 to 100 in increments of 5.

4. Run PSpice, select **All** available sections, and generate the default graph.

5. Switch to performance analysis (**Plot**, **X-axis Settings**, **Performance Analysis**, **OK**) and note that the X-axis switches from frequency to Rtank (RVAL) values from 1 (0) to 100.

6. Command the *bandwidth* goal function to search each impedance curve and return the bandwidth for display as follows: **Trace**, **Add**, **bandwidth(1)**, enter **V(Vsource:+)/I(Vsource)** within the parentheses, **OK**. The resulting performance analysis plot is shown in Figure 27.3.

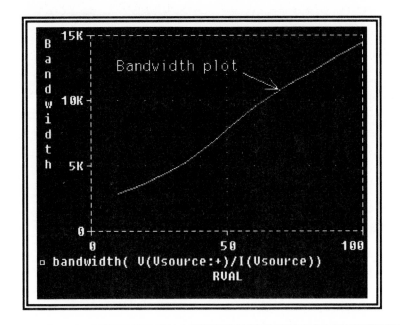

FIGURE 27.3

Plot of bandwidth versus Rtank

7. Based on the results of Figure 27.3, what value of Rtank (rounded off to the nearest whole number) will give our radio station a bandwidth of almost exactly 8kHz?

    Rtank (@BW of 8kHz) = _____

8. Add a second Y-axis to the graph and include a performance analysis plot of impedance *amplitude* versus Rtank. The result is shown in Figure 27.4.

    Is the peak impedance amplitude at a bandwidth of 8kHz between the values of 1kΩ and 3kΩ, as required?

    Yes   **No**

PSpice for Windows

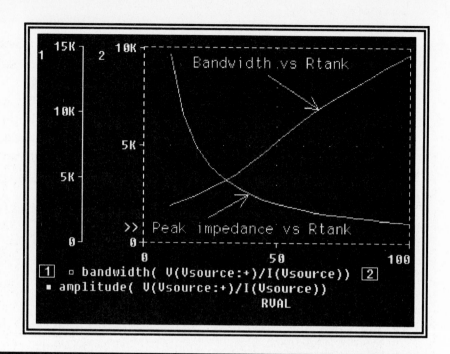

FIGURE 27.4

Adding a plot of
peak impedance

Histogram

9. To prepare for a Monte Carlo analysis, replace part R (attribute Rtank) with part Rbreak, give the new part the same attribute, and assign it the value determined in step 8. (Remember that Rbreak is both the part and model name, and only the model name is displayed on the schematic.)

10. Select Rtank, enter the model Editor (**Edit**, **Model**, **Edit Instance Model**), and enter *R = 1 DEV = 10% and LOT = 2%*, **OK**.

11. Disable the parametric analysis and enable Monte Carlo/Worst Case analysis. Within the Monte Carlo dialog box, set *Analysis* to **Monte Carlo**, choose 100 MC runs, set *Analysis Type* to **AC**, and *Output* to **All** (under MC Options). [Because we will not be accessing the output file, set *Output Var* and *Function* to any convenient terms, such as I(Vsource) and YMAX.]

12. Run PSpice, select **All** in the *Available Sections* box, **OK**, and note the default probe graph.

13. Switch to performance analysis (**Plot**, **X-axis Settings**, **Performance Analysis**, **OK**). Note the initial histogram chart.

14. Generate the bandwidth histogram of Figure 27.5. (**Trace**, **Add**, **bandwidth(1)**, enter **V(Vsource:+)/I(Vsource)** in parentheses, **OK**). As shown, the X-axis is bandwidth and the Y-axis is percent. Also shown is the wealth of statistical data generated.

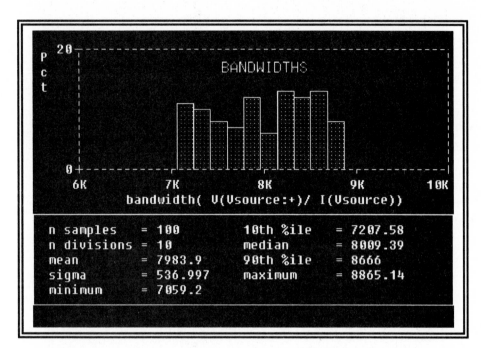

FIGURE 27.5

Bandwidth histogram

15. Based on the random (Monte Carlo) tolerance variations of Rtank about its correct value, the histogram shows the percentage of time the bandwidth will fall into one of 10 "slots." Considering the results of Figure 27.5:

    (a) What bandwidth slot or slots (1 to 10, left to right) is most and least likely?

       Most likely = _____   Least likely = _____

    (b) What is the minimum, maximum, and average (mean) expected bandwidth values?

       Min = _____   Max = _____   Mean = _____

    (c) *Approximately* what percentage of the time would you expect the bandwidth to be above 8500Hz?

       % time above 8500Hz = _____

PSpice for Windows

Advanced Activities

16. Continuing from step 15, follow the steps below to change the tolerance variation from the default *uniform* case to the more realistic *Gaussian* distribution.

    Bring up the Rbreak model (select Rtank, **Edit**, **Model**, **Edit Instance Model**, etc.), change model name Rbreak to any other name (such as Rbreak2), change DEV to DEV/GAUSS and LOT to LOT/GAUSS, **OK**, and generate another histogram.

    Looking at the result (Figure 27.6), we see a more realistic histogram. This time, *approximately* what percentage of the time would you expect the bandwidth to be above 8500Hz? [Compare to step 15(c).]

    % time above 8500Hz (Gaussian) = _____

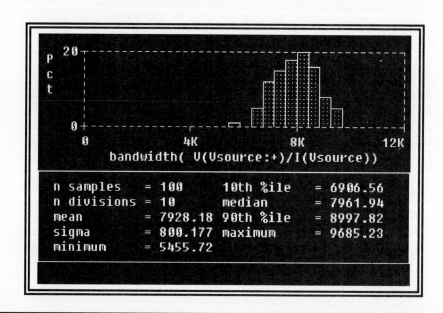

FIGURE 27.6

Histogram using
Gaussian tolerance distribution

17. To help us "fine tune" the resonant frequency of our radio station, make Ctank a parameter (Figure 27.7), write a goal function to return the resonant frequency, and generate the performance analysis graph of Figure 27.8. What value of Ctank will give a resonant frequency of exactly 50kHz?

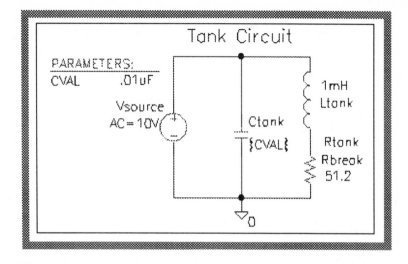

FIGURE 27.7

Circuit with varying
Ctank

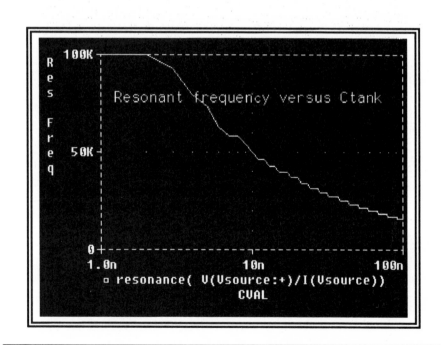

FIGURE 27.8

Plot of resonant
frequency versus
Ctank

EXERCISES

- Using the automotive suspension example of Chapter 7, generate performance analysis traces of *rise time* and *overshoot* versus Rshock.

  [Goal function *rise time* will return the time it takes the initial curve to go from 10% to 90% of maximum; goal function *overshoot* will make use of the XVAL command to return the percentage difference between the maximum amplitude and the steady-state amplitude. Also, *search forward XVAL(2.0)!2* will generate the steady-state amplitude (y2) at 2 seconds.]

  (Hint: File *msim.gf* contains a number of commonly used goal functions.)

QUESTIONS & PROBLEMS

1. Does performance analysis require multiple runs of a waveform?

   Yes   **No**

2. When using performance analysis, the X-axis is

   (a) the single returned value from the goal function.
   (b) the varying parameter value.

3. A goal function returns how many values when it analyzes a single waveform?

4. Write a goal function that returns the time that the *second* peak occurs in a waveform (such as the damped sine wave waveforms of Chapter 7).   (Hint: Review Probe Note 9.1.)

5. When a performance analysis is run on a family of curves generated by the random tolerance variation of a Monte Carlo analysis, we generate a

    _____.

6. The X-axis of a histogram is

    (a) the returned goal function values.
    (b) percentage.
    (c) time or frequency.

7. How can we change the number of histogram slots (presently ten)?   (Hint: **Tools**, **Options** under *Probe*.)

8. Why is a Gaussian-based histogram (Figure 27.6) usually more realistic than a conventional histogram (Figure 27.5)?

CHAPTER 28

Controlled Sources
Analog Behavioral Modeling

OBJECTIVES

- To use *controlled sources* to model a circuit element.
- To add *behavioral modeling* to simulate more complex circuit components.

DISCUSSION

CONTROLLED SOURCES

To simplify initial circuit design and to model customized components, PSpice offers the four types of *controlled sources* listed below:

Device	Description
E	VCVS (Voltage-Controlled Voltage Source)
F	ICIS (Current-Controlled Current Source)
G	VCIS (Voltage-Controlled Current Source)
H	ICVS (Current-Controlled Voltage Source)

These devices have ideal input/output characteristics and simple transfer functions. For example, Figure 28.1 shows an E device (a VCVS) used as a simple voltage amplifier with a gain of 10. By generating various time- and frequency-domain test curves, such as those in Figure 28.2, we find that it is a "perfect" voltage amplifier, with a gain of exactly 10, infinite Z_{in}, zero Z_{out}, infinite bandwidth, zero harmonic distortion, and totally noiseless.

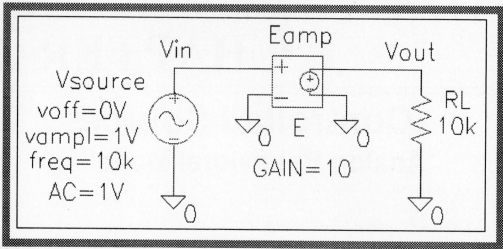

FIGURE 28.1

The E device as a
voltage amplifier

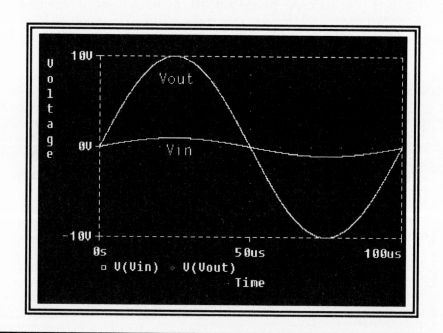

FIGURE 28.2

E device V_{in} and
V_{out} waveforms

BEHAVIORAL MODELING EXTENSIONS

Because the simple E, F, G, and H devices are so perfect, they are not
always appropriate in a real-world environment. For this reason,
PSpice also offers special extended versions of the E and G devices
that provide *analog behavioral modeling*.

   Analog behavioral modeling allows the designer to specify complex transfer functions that more closely represent actual circuit components and systems.   For each of these special E and G type controlled sources we can select from among the following mathematical relationships:

- **Sum (SUM)**
- **Multiply (MULT)**
- **Table (TABLE)**
- **Value (VALUE)**
- **Frequency (FREQ)**
- **Laplace (LAPLACE)**
- **Chebyshev Filters (CHEBYSHEV)**

   As an example of the use of these extensions, let's model a single JFET for use in a voltage amplifier. Since a JFET (when properly biased) is a VCIS with a complex transfer function, we select the GVALUE model. We specify the parabolic transfer function from Chapter 18, add a load resistor and bias voltage, and generate the amplifier circuit of Figure 28.3.

   When we test the amplifier, it does give the expected results (Figure 28.4). (Be aware that the GJFET's other characteristics, such as Z_{in}, Z_{out}, bandwidth, and noise, are still perfect. However, as expected, it does show a total harmonic distortion of approximately 6%.)

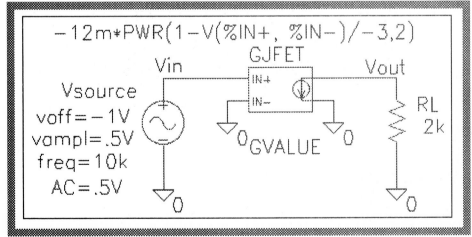

Figure 28.3

Behavioral modeling
of a JFET amplifier

PSpice for Windows

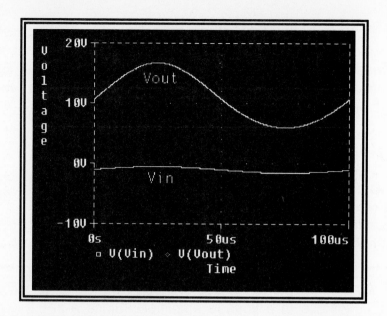

FIGURE 28.4

JFET model
waveforms

FREQUENCY-DOMAIN MODELING

The frequency-domain models (FREQ and LAPLACE) are considerably more complex because the output is not instantaneous with each input value, but depends on the input characteristics over time (such as its frequency).

To model a filter, such as the simple passive low-pass filter of Figure 28.5, we first calculate several of its characteristics:

- **Fbreak = 1 / $2\pi RC$**
- **Rolloff = 20dB/dec   A(low frequency) = 0dB**
- **Phase = atan(X_C/R)**

We then use these characteristics to form a look-up table to describe the frequency response and to program the device.

Frequency	A(dB)	Phase
0	0	0
160	-3	-45
1.6k	-20	-84
16k	-40	-89
160k	-60	-90

PSpice for Windows

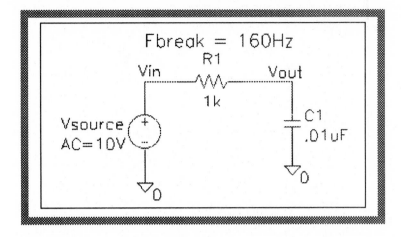

FIGURE 28.5

A passive
low-pass filter

Next, during our *simulation practice*, we recreate the circuits of Figures 28.1, 28.3, and 28.5 and verify that they have the expected properties.

SIMULATION PRACTICE

The VCVS Device

1. Draw the VCVS amplifier of Figure 28.1 and set the attributes as shown.   (<u>Note</u>: The E, F, G, and H controlled source devices are found in library *analog.slb*. **DCLICKL** on the E device to set gain.)

2. Using conventional techniques, generate the curves of Figure 28.2. By examining the results of these curves, *and others as needed* (i.e., AC sweep), determine each of the following:

    A   = _____
    Z_{in} = _____
    Z_{out} = _____
    BW   = _____
    HD   = _____
    ONOISE = _____

3. Based on the results of Step 2, does the E device have "perfect" VCVS characteristics?

    Yes   **No**

Behavioral Modeling

4. Draw the circuit of Figure 28.3 and set all the attributes as shown —except the transfer function. (The GVALUE VCIS device is in library *abm.slb*.)

5. To set the VCIS characteristics of device GJFET, we review Chapter 18 and obtain the transconductance characteristics for the J2N3819 device we wish to model:

    $$I_D = I_{DSS}(1 - V_{GS}/V_{GSoff})^2$$

    From Chapter 18, we find that I_{DSS} = 12mA and V_{GSoff} = -3V. Using these values, verify that the following equation is correct:

    $$I_D = 12mA(1 - V_{GS}/-3)^2$$

6. **DCLICKL** on GJFET device to bring up its *Part Name* dialog box and enter the following in the *value* box:

    –12m*PWR(1 – V(%IN+,%IN–)/–3,2).

    where *V(%IN+,%IN-)* refers to the voltage between the two input nodes. (The minus sign is required to simulate the 180° phase shift.)

7. Set up the system for transient and Fourier analysis (DFT). Run PSpice and generate the curves of Figure 28.4. What is the approximate (average) gain of the amplifier?

    A = _____

8. Locate the Fourier section of the output file and record the total harmonic distortion.

    HD = _____

9. Using the DC sweep mode, generate the transconductance curve of Figure 28.6 and compare to those of Chapter 18. Does the curve match the equations of step 5? (<u>Hint</u>: A DC sweep can be performed with source VSIN.)

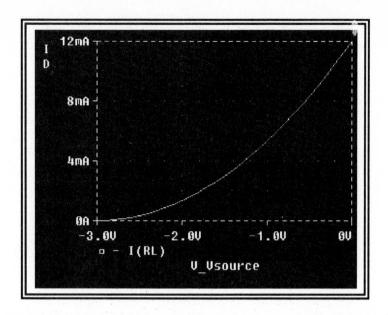

FIGURE 28.6

The GJFET
transconductance
curve

Advanced Activities

10. To model the low-pass filter of Figure 28.5, draw the circuit of Figure 28.7—which uses the EFREQ device (from library *abm.slb*).

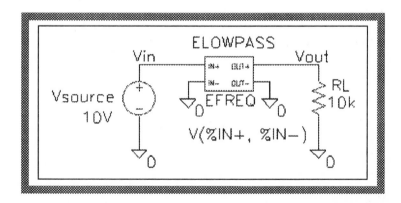

FIGURE 28.7

Modeling a
low-pass filter

11. Bring up the *Part Name* dialog box for part EFREQ (**DCLICKL** on part), **CLICKL** on *TABLE*, and enter the following in the *Value* box (see the discussion for an explanation of these values):

    (0,0,0)(160,-3,-45)(1.6k,-20,-84)(16k,-40,-89)(160k,-60,-90)

12. Perform an AC sweep and generate the Bode plot of Figure 28.8. Does the device reasonably model a low-pass filter?

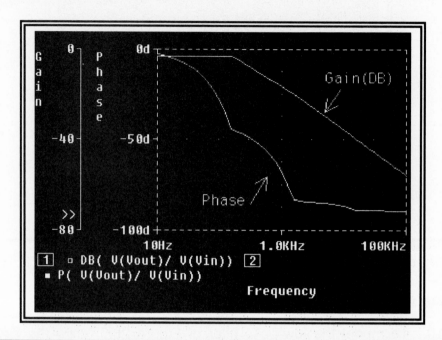

FIGURE 28.8

Bode plot of modeled
low-pass filter

13. Enter a square wave of approximately 100Hz (PER = 10ms) into the low-pass filter model of Figure 28.7, perform a transient analysis, and note the result. Does the low-pass filter round off the sharp edges?

Generate an FFT for both the input and output waveforms. Does the filter greatly reduce the harmonic content of the input waveform?

EXERCISES

- Design and test a behavioral model for the tank circuit of Chapter 4. (Hint: Use a table.)

- Design and test a behavioral model for a bipolar transistor. Can you generate collector curves?

- Determine the total harmonic distortion (DFT) of both the input and output signals of the low-pass filter of Figure 28.7.

QUESTIONS & PROBLEMS

1. What are the major characteristics of a VCVS controlled source (E device)? Of an ICIS controlled source (F device)?

2. What do you think is the relationship between input and output of the GMULT device?

3. Why should the output JFET waveform of Figure 28.4 show harmonic distortion?

4. Why is frequency-domain analysis (using a FREQ device) referred to as *noninstantaneous?*

5. Based on the E and GValue devices studied in this chapter, would you expect the EFREQ device of Figure 28.7 to also have perfect Z_{in} and Z_{out} characteristics (infinity and zero)?

CHAPTER 29

Modular Design
Hierarchy

OBJECTIVES

- To demonstrate the concepts of *top-down* design.
- To create a circuit that is composed of levels of hierarchy.

DISCUSSION

The one word that separates today's circuits from those of the past is *complexity*. Clearly, new techniques must be used when we move from a small-scale circuit of 20 components to a large-scale circuit of thousands of components.

The solution to working in any complex environment is *top-down* design. The key element in top-down design is *modularization*, in which a large-scale, complex task is broken down into a hierarchy of modules, from the general and conceptual at the top, to the specific and detailed at the bottom.

Modularizing a program into a hierarchy of modules yields a number of benefits:

- We can focus our attention on a single module at a time, without being hindered by the complexities of the entire circuit.
- Our initial design is from a high-level perspective, in which concepts are important and low-level details can be ignored.
- We are encouraged to create customized low-level "tools" that can be stored in a library and used over and over (so we don't spend our time "reinventing the wheel").

The major components of top-down design are *blocks, hierarchical parts, primitive parts*, and *views*. A block is a rectangular "black box" that represents a collection of circuitry, a hierarchical part is a nested block within another block, a view allows a block to have more than one solution, and a primitive part is at the lowest nesting level and contains only circuit elements.

In this chapter, we design an amplifier using the techniques of top-down design.

SIMULATION PRACTICE

1. We begin our design at the top. All we know at this time is that there is a source, a load, and a DC power supply; the amplifier itself is a block.

   Based on these initial parameters, draw the top-level design of Figure 29.1, set the attributes as shown, and save the circuit using the suggested file name *top*.

   > The block is created by **Draw Block**. To change the size of the block, hold shift down, **CLiCKRH** (click right and hold down), and drag the sides of the block to the desired shape. When wires are drawn to the edges of the block, the pin numbers appear automatically *in the numerical order in which they are drawn*.)

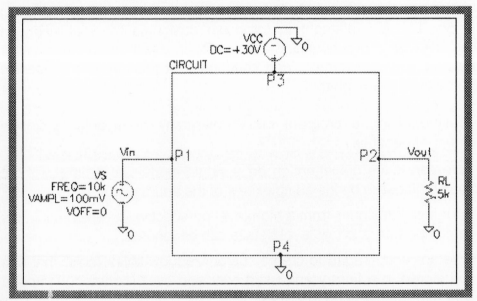

FIGURE 29.1

The top-level design

2. The next step is to *push* inside the top-level circuit block in order
    to create the middle-level design.   To accomplish this: **CLICKL** to
    select the block, **Navigate**, **Push** (or simply **DCLICKL** within the
    block) and bring up the *Setup Block* dialog box.   For this middle-
    level circuit, enter the suggested file name *middle*, **OK**. (Note the
    four *interface ports*, whose numbers correspond to the *top* block.)

3. We are now faced with a crucial design question: What overall
    circuitry do we need to create an amplifier?   Because of the small
    value of the load (500Ω), we select a two-stage amplifier/buffer.

    Using the interface ports, create the midlevel, more detailed
    design of Figure 29.2.   (Reminder: **DCLICKL** on any attribute,
    including the pin numbers, to change or increase in size.)

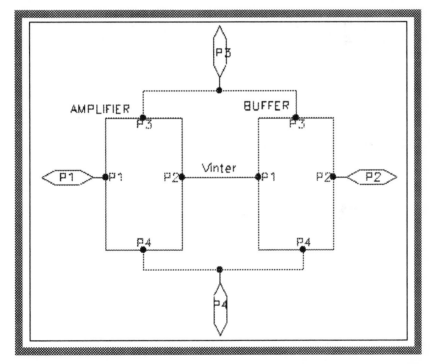

FIGURE 29.2

Midlevel design

4. Next, we must *push* into each middle-level block in order to create
    low-level circuitry.   We start with the amplifier block: select,
    Navigate, **Push** (or **DCLICKL** within block *AMPLIFIER*), and enter
    the suggested file name *lowamp* in the *Setup* dialog box, **OK**.

5. We are now at the lowest level, and it is time to design the amplifier. Because we require high gain, and because we can tolerate some degree of harmonic distortion, we choose a low-cost bipolar design, using collector-feedback biasing.

    Draw the *primitive part* of Figure 29.3, again being careful to place the interface ports at the correctly numbered locations.

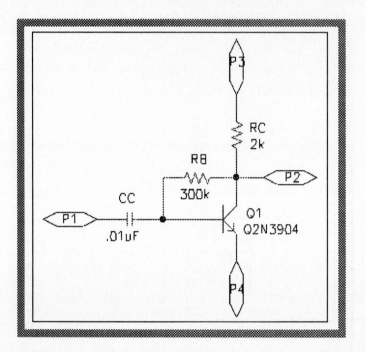

FIGURE 29.3

Primitive part
Lowamp

6. Return to the middle-level design (**Navigate**, **Pop**, Save Changes).   (*Navigate pop* returns to the next highest level, *navigate top* returns to the highest level.)

7. The last step is to design the buffer stage. This time, to show how *previous* circuits can be used as hierarchical parts, we will place a circuit we have already designed into block *BUFFER*. Because our circuit is battery-powered, we require a design of high efficiency.

    Review the highly efficient Class B buffer of Figure 29.4, first introduced in Chapter 17.

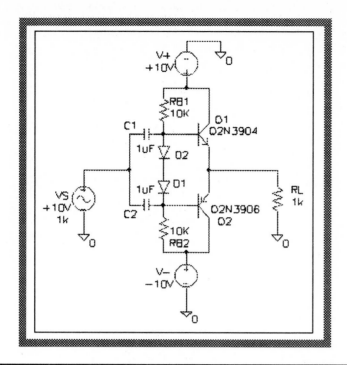

FIGURE 29.4

Previously designed
Class B buffer
from Chapter 17

8. Select the BUFFER block, **Navigate**, **Push** to open the *Setup Block* dialog box, and enter the full file name (including directory, if different) of the Class B buffer of Figure 29.4. (Note: If the Class B buffer of Figure 29.4 does not exist, leave the present design, draw the circuit under a new file name, and return to the beginning of this step.)

9. Modify the circuit to match the single power supply and interface port needs of Figure 29.5. [This time, because the circuit already exists, we must obtain the *INTERFACE* ports from the *port.slb* library (**DCLICKL** to assign port numbers).]

10. Our design is now finished. To test the system, return to the top level (**Navigate**, **Top**), and set up PSpice for a transient analysis from 0 to .2ms.

11. Generate the input/output curves of Figure 29.6. What is the gain of the overall circuit?   (Is this reasonable?)

    A = _____

PSpice for Windows

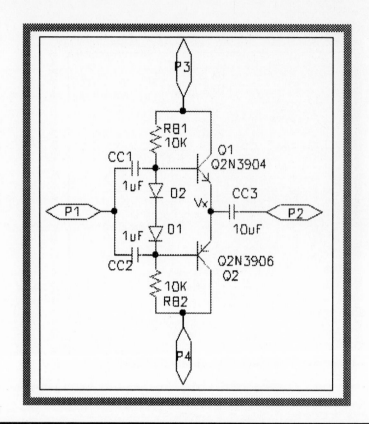

FIGURE 29.5

Previous design
modified for
hierarchical placement

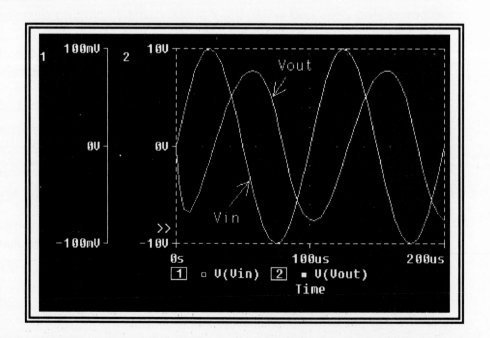

FIGURE 29.6

Circuit waveforms

PSpice for Windows

12. To show that we can display any waveform at any level from Probe, display the following on the appropriate plot: (Note the use of "dots" to show the node level.)

    (a) Voltage node *Vinter* from the middle-level block [**Trace**, **Add**, V(CIRCUIT.Vinter)].

    (b) The low-level buffer voltage at the emitter of Q2 [**Trace**, **Add**, V(CIRCUIT.BUFFER.Q2:e)].

13. Perform a FFT (fast Fourier transform) on the output waveform (**Plot**, **X-axis**, **Fourier**, **OK**). Is the harmonic distortion reasonably low?

    Yes   **No**

Navigation Path

14. Using PSpice, we have the ability to determine at any time exactly where we are in the hierarchy of modules.

    For example, push into *BUFFER*, **Navigate**, **Where** to open up the *Where* dialog box. *Current Schematic* shows the current file name level, *Hierarchy Path* lists the sequence of block names that we pushed into, and *Levels Pushed Into* lists the same path in terms of file names.

Advanced Activities (Views)

15. During the design of a complex circuit, it is often desirable to give certain blocks alternative solutions, called *views*.

    As an example of a view, we may wish to assign block *BUFFER* a temporary solution while we complete the rest of the design. An ideal temporary solution is a *controlled source* from Chapter 28.

    To create this second view, navigate to the middle-level circuit (Figure 29.2).

16. Select hierarchical part *BUFFER*, **Edit**, **Views** to bring up the Block *BUFFER* dialog box, enter any desired *View Name* (such as CS, for *Controlled-Source)* and *Schematic Name* (such as VCVS), **Save View**, **OK**.

17. Block *BUFFER* now has two views (CS and default). To create the schematic circuit for view CS, select *BUFFER*, **Navigate**, **Push** to bring up the *Select View* dialog box, **CLICKL** on *CS=VCVS* to select this view, **OK**, **Yes** on *Save Current Changes*, and bring up the schematic block with the expected four interface ports.

18. Draw the circuit of Figure 29.7 and set the attributes as shown. Controlled source *E* is from library *analog.slb*. (Note that interface port 3 is not needed and can be left "floating.")

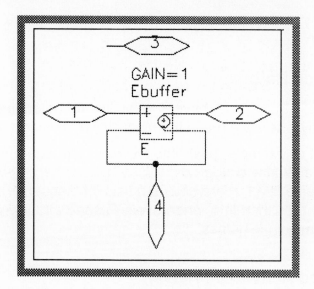

FIGURE 29.7

Alternative schematic
view VCVS

19. The second view for *BUFFER* finished, we return to the middle level (**Navigate**, **Pop** to save current changes).

20. To show how the circuit designer can bring up either view of block *BUFFER* during schematic design, push into *BUFFER* and select either *lowbuf* (the default view) or *VCVS*. (To see the views that are available with any block, select the block and **Edit**, **Views**, note the listed views, **OK** when done.)

21. To command PSpice to select view CS during netlisting: **Options**, **Translators**, select PSpice (under *Translator*), select CS (under *View*), **Apply**, **OK**.

22.   Return to the top level (**Navigate**, **Top**) and Run PSpice.

23.   Display the input and output of block *BUFFER* [V(CIRCUIT.Vinter) and V(Vout)].   Are these the waveforms you would expect from a "perfect" buffer? (Hint: Are the two waveforms *exactly* the same?)

   Yes   **No**

EXERCISES

- Design the power supply of Chapter 9 using the techniques of hierarchy.

- Design a three-stage audio amplifier (FET input, bipolar middle, Class B output) using the techniques of hierarchy.

QUESTIONS & PROBLEMS

1.   When using top-down design, what kinds of modules are at the top and what kind are at the bottom?

2.   What are the advantages of modularization?

3.   What is the difference between a *hierarchical part* and a *primitive part*?

4.   With PSpice, how many levels of hierarchy are possible?

5.   To go from a higher level module to a lower level module we

   (a) Push.
   (b) Pop.

6.   What is displayed by **Navigate**, **Where**?

7.   Can more that one circuit belong to a given module?   (Hint: What is a view?)

PART VII
Analog Communications

In the single chapter of Part VII, we concentrate on analog communications.

Historically, the first form of analog communications was called *amplitude modulation*, the subject of Chapter 30. We will find that the processes of *modulation* and *detection* can be performed by the quite ordinary components and circuits introduced in previous chapters.

CHAPTER 30

Amplitude Modulation
Detection

OBJECTIVES

- To generate an amplitude-modulated (AM) signal.
- To detect an AM signal.

DISCUSSION

An electromagnetic wave is generated whenever an electron is accelerated. However, transmission of electromagnetic waves through the atmosphere is only efficient at frequencies well above audio. This led to the concept of a low-frequency information-carrying signal *modulating* a high-frequency *carrier*. The first form of modulation was *amplitude modulation*.

AMPLITUDE MODULATION

A signal is amplitude modulated when a low-frequency information signal controls the amplitude of a high-frequency carrier. A bipolar transistor can be used to modulate a signal because its gain depends on bias current (re' = $25mV/I_{EQ}$).

The term *percent modulation* is a measure of the strength of the modulating signal. It is defined as follows:

$$\% \text{ Modulation} = \frac{\text{Maximum gain} - \text{Minimum gain}}{\text{Nominal gain (no modulation signal)}} \times 100$$

In the frequency domain, the modulated signal contains *sidebands* equal to the sum and difference of the carrier and modulation signals. A Fourier analysis of the output signal will reveal these information-carrying sidebands.

When the modulated signal is received, it must be *demodulated* (the information signal extracted from the carrier). The simplest demodulator is a peak rectifier. The RC time constant must be short compared with the modulating period, but long compared to the carrier period.

SIMULATION PRACTICE

1. Draw the modulation circuit of Figure 30.1. (<u>Note</u>: A lower-than-normal 100kHz carrier frequency is used to shorten computation times.)

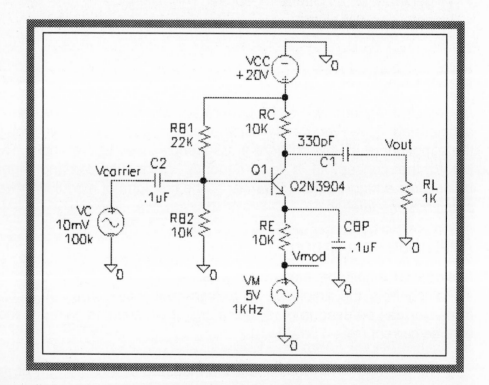

FIGURE 30.1

Amplitude modulation circuit

2. Generate the output waveforms of Figure 30.2 and determine the percent modulation.

Percent modulation = _____

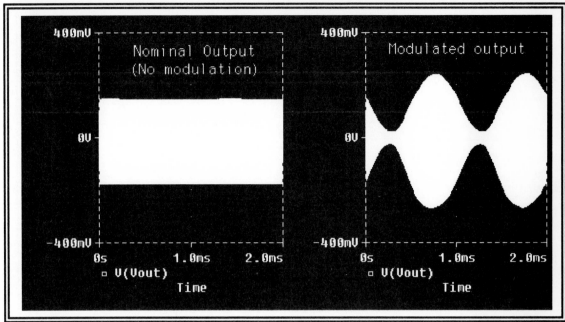

FIGURE 30.2

Unmodulated versus modulated output

3. Reduce the audio input amplitude (V_M) to 2V, generate V_{out}, and determine the reduced percent modulation.

   Percent modulation (V_M = 2V) = _____

Demodulation (Detection)

4. Add the amplifier/detector to your modulation circuit (Figure 30.3).

5. If the circuit works as expected, what should be the shape and frequency of the output signal ($V_{receive}$)?

   Expected shape = _____

   Expected frequency = _____

6. Place an arrow at the point in the circuit where the signal would normally be transmitting into the atmosphere by way of an antenna.

7. Create the plots of Figure 30.4, which compare the modulation, carrier, transmit, and receive waveforms. Does $V_{receive}$ have the expected shape and frequency?

   Yes   No

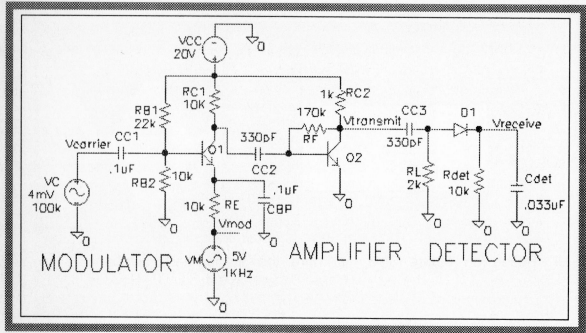

FIGURE 30.3

Adding AM detection

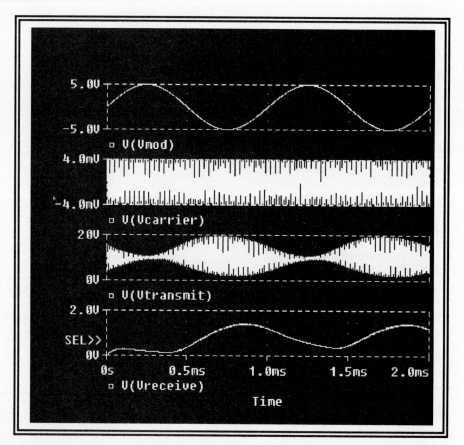

FIGURE 30.4

Modulation, carrier, transmit,
and receive waveforms

PSpice for Windows

Fourier Analysis

8. Use FFT (fast Fourier transform) analysis on the time-domain curves of Figure 30.4 (and expand the X-axis) to generate the frequency spectrum plots of Figure 30.5. (From Probe, **Plot**, **X-Axis Settings**, **Fourier**, **OK**.)

9. The frequency spectrum waveforms of Figure 30.5 are hard to see because appropriate X-axis ranges are inconsistent. To solve the problem, generate the plots of Figure 30.6, which use a combination of multiple Y-axes, multiple plots, unsynched X-axes (*Probe Note 11.1*), and user-defined X- and Y-axes settings.

10. Based on Figure 30.6, list here the "transmitted" sideband frequencies (at the peak of each Vtransmit sideband). How do the frequencies relate to the modulation frequency?

    f (lower sideband) = _____

    f (upper sideband) = _____

11. Is the only major difference between V_{mod} and $V_{receive}$ the presence of a DC component in $V_{receive}$?

    Yes   No

Advanced Activities

12. Using hand calculations on the circuit of Figure 30.1, determine the instantaneous gain at the values of V_{mod} listed here. Using your data, determine the percent modulation of the output. (<u>Hint</u>: Voltage gain = $R_C \| R_L / re'$ and re' = 25mV/I_E.)

    Nominal gain (V_{mod} = 0V) = _____

    Maximum gain (V_{mod} = -5V) = _____

    Minimum gain (V_{mod} = +5V) = _____

    Percent modulation = _____

13. By examining waveforms before and after capacitor CC2 (Figure 30.1 or 30.3), what is the purpose of CC2? (<u>Hint</u>: What happened to the 1kHz modulation signal?)

14. Reduce the percent modulation (Figure 30.3) and compare the resulting frequency spectrum with Figure 30.6.   Explain the difference.

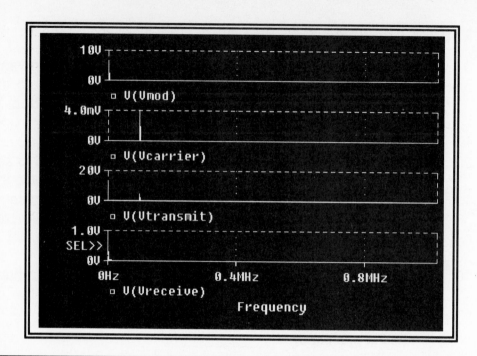

FIGURE 30.5

Frequency spectrum
waveforms

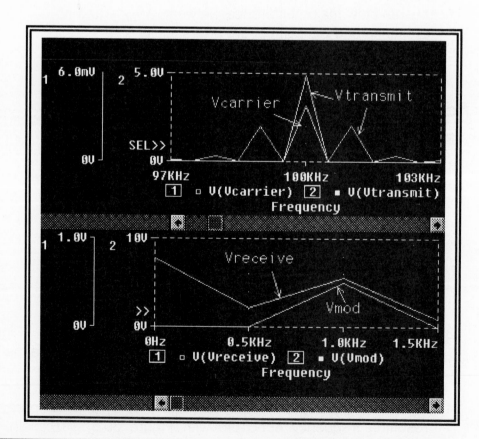

FIGURE 30.6

Reformatted waveforms

PSpice for Windows

15. Modulate the system with a square wave and compare the frequency spectrum of the input and output. What does the result say about the harmonic content of a square wave and the ability of the circuit to pass high modulation frequencies?

16. Modulate the system with a *damped* sine wave and view the results in the time domain.

EXERCISES

• Design an AM modulation and detection system that matches a specific station within the AM band (such as 740kHz).

QUESTIONS & PROBLEMS

1. Which of the following circuit elements is responsible for the gain variation during amplitude modulation?:

    (a) RL
    (b) CC1
    (c) re'
    (d) VCC

2. Capacitor CBP is designed to short which of the following signals to ground?:

    (a) Vcarrier
    (b) Vmodulation

3. Why must the key element of a modulation circuit (such as the bipolar transistor of Figure 30.1) be a *nonlinear* device?

4. What is the purpose of the *carrier* signal?

5.   What is the difference between a *detector* and a *peak rectifier*?

6.   For a real broadcast, why would the Fourier spectrum of the transmitted signal likely be a continuum?

APPENDIX A

Notes

Schematics

PSpice

3.1   What controls do we have over the transient analysis process?
3.2   How does PSpice perform transient calculations?

Probe

2.1   How do I enter custom Y-axis variables?
2.3   How do I print my circuit, graph, or output file?
3.1   How do I use the cursor to accurately determine waveform values?
3.2   How do I create multiple Y-axes?
3.3   How do I place legend symbols on the curves?
4.1   What special markers are available to plot advanced waveforms?
4.2   How do I document my Probe graphs?
5.1   How do I expand or compress waveforms?
6.1   How do I mark coordinate values on my graphs?
6.2   How do I generate multiple plots within a single Probe window?
6.3   How do I label the Y-axis?
7.1   How do I single out individual curves from a family of curves?
8.1   How do I change the X-axis variable?
9.1   Does Probe include any "tricks" to help measure ripple and other waveform parameters?
11.1   How do I uncouple individual plots of a multiple-plot graph?

APPENDIX B

Probe's Mathematical Operators

Probe Function	Description
( )	Grouping
-	Logical complement
*/	Multiply/divide
+−	Add/subtract
&^|	AND, Exclusive OR, OR
AVGX(x,d)	Average (x to d)
RMS(x)	RMS average
DB(x)	x in dB
MIN(x)	Minimum real part of x
MAX(x)	Maximum real part of x
ABS(x)	Absolute value of x
SGN(x)	+1 if X>0, 0 if x = 0, -1 if x<0
SQRT(x)	$X^{1/2}$
EXP(x)	e^x
LOG(x)	Ln(x)
LOG10(x)	log(x)
M(x)	Magnitude of x
P(x)	Phase of x (degrees)
R(x)	Real part of x
IMG(x)	Imaginary part of x
G(x)	Group delay of x (sec)
PWR(x,y)	$(x)^y$
SIN(x)	sin(x)
COS(x)	cos(x)
TAN(x)	tan(x)
ATAN(x)	$\tan^{-1}(x)$
d(x)	Derivative of x with X-axis
s(x)	Integral of x with X-axis
AVG(x)	Average of x

APPENDIX C

Scale Suffixes

Symbol	Scale	Name
F	10E−15	*femto-*
P	10E−12	*pico-*
N	10E−9	*nano-*
U	10E−6	*micro-*
M	10E−3	*milli-*
K	10E+3	*kilo-*
MEG	10E+6	*mega-*
G	10E+9	*giga-*
T	10E+12	*tera-*

APPENDIX D

Spec Sheets

D1N750
ZENER DIODE

Rating	Symbol	Value	Unit
DC Power Dissipation @ TA <= 50 $^{\circ}$C Derate > 50°C	PD	500 3.3	mW mW/$^{\circ}$C
Operating and Storage Junction Temperature Range	TJ, Tstj	-65 to +200	$^{\circ}$C

Type Number	Nominal Zener Voltage VZ@IZT Volts	Test Current IZT mA	Maximum Zener Z Zzt@Izt Ohms	Maximum Zener Current IZM mA	TA = 25°C IR @ VR = 1V uA	TA = 150°C IR@VR = 1V uA
D1N750	4.7	20	19	75 95	2	30

2N3904/3906
NPN/PNP SILICON SWITCHING & AMPLIFIER TRANSISTORS

Rating	Symbol	Value	Unit
Collector-Base Voltage	VCB	60	Vdc
Collector-Emitter Voltage	VCEO	40	Vdc
Emitter-Base Voltage	VEB	6.0	Vdc
Collector current	IC	200	mAdc
Total Power Dissipation @ TA = 60°C	PD	250	mW
Total Power Dissipation @ TA = 25°C	PD	350	mW
Derate above 25°C		2.8	mW/$^{\circ}$C
Total Power Dissipation @ TC = 25°C	PD	1.0	mW
Derate above 25°C	PD	8.0	mW/$^{\circ}$C
Junction Operating Temperature	TJ	150	$^{\circ}$C
Storage Temperature Range	Tstg	-55-+150	$^{\circ}$C
Characteristic	**Symbol**	**Max**	**Unit**
Thermal Resistance, Junction to Ambient	R0jA	357	$^{\circ}$C/W
Thermal Resistance, Junction to Case	R0jC	125	$^{\circ}$C/W

Characteristic	Symbol	Min	Max	Unit
Collector-Base Breakdown Voltage (IC = 10uAdc, IE = 0)	BVcbo	60		Vdc
Collector-Emitter Breakdown Voltage (IC = 1mAdc, IB = 0)	BVceo	60		Vdc
Emitter-Base Breakdown Voltage (IE = 10uAdc, IC = 0)	BVebo	6.0		Vdc
Collector Cutoff Current (VCE = 30 Vdc, VEB(off) = 3.0 Vdc	Icex		50	nAdc
Base Cutoff Current (VCE = 30 Vdc, VEB(off) = 3.0 Vdc	Ibl		50	nAdc
DC Current Gain (IC = 0.1 mAdc, VCE = 1.0 Vdc) (IC = 1.0 mAdc, VCE = 1.0 Vdc) (IC = 10 mAdc, VCE = 1.0 Vdc) (IC = 50 mAdc, VCE = 1.0 Vdc) (IC = 100 mAdc, VCE = 1.0 Vdc)	Hfe	40 70 100 60 30	300	
Collector-Emitter Saturation Voltage (IC = 10 mAdc, IB = 1.0 mAdc) (IC = 50 mAdc, IB = 5.0 mAdc)	VCE(sat)		.2 .3	Vdc
Base-Emitter Saturation Voltage (IC = 10 mAdc, IB = 1.0 mAdc) (IC = 50 mAdc, IB = 5.0 mAdc)	VBE(sat)	.65	.85 .95	Vdc

2N5484-5486 (Similar to 2N3819)
JFET
Maximum Ratings

Rating	Symbol	Value	Units
Drain-gate voltage	VDG	25	Vdc
Reverse gate-source voltage	VGSR	25	Vdc
Drain Current	ID	30	mAdc
Forward Gate Current	IG(f)	10	mAdc
Total Device Dissipation @ TC = 25°C Derate above 25°C	PD	310 2.82	mW mW/°C
Operating and Storage Junction Temperature Range	Tj, Tstg	-65 to +150	°C

ELECTRICAL CHARACTERISTICS

Characteristic	Symbol	Min	Typ	Max	Unit
Gate-Source Breakdown Voltage (IG = -1.0uAdc, VDS = 0)	V(BR)GSS	-25			Vdc
Gate Reverse Current (VGS = -20 Vdc, VDS = 0)	IGSS			-1.0	µAdc
Gate-Source Cutoff Voltage (VDS = 15Vdc, ID = 10nAdc)	VGS(off)	-.5		-4.0	Vdc
Zero-gate-voltage Drain Current (VDS = 15Vdc, ID = 10nAdc)	IDSS	4.0		10	mAdc
Forward Transfer Admittance (VDS = 15Vdc, VGS = 0, f = 1kHz)	Yfs	3500		7000	µmhos
Input Admittance (VDC = 15Vdc, VGS = 0, f = 100 MHz.)	Re(Yis)			100	µmhos
Output Admittance (VDS = 15Vdc, VGS = 0, f = 1.0MHz.)	Yos			60	µmhos
Output Conductance (VDS = 15Vdc, VGS = 0, f = 100MHz.)	Re(Yos)			75	µmhos
Forward Transconductance (VDS = 15Vdc, VGS = 0, f = 100MHz.)	Re(Yfs)	3000			µmhos

APPENDIX E

Windows Tutorial

The following conventions are used within the text and this tutorial.

CLICKL or **BOLD PRINT** (*click left once*) to select an item
DCLICKL (double *click left*) to end a mode or edit a selection.
CLICKR (*click right once*) to abort a mode.
DCLICKR (double *click right*) to repeat an action.
CLICKLH (*click left, hold down and move mouse*) to drag a
        selected item. Release left button when placed.
DRAG (*no clicks, move mouse*) to move an item.

Follow the steps below to review the use of the mouse, changing the size and location of windows, pull down menus, and dialog boxes.

1. Turn on the computer and bring up the *Program Manager* window of Figure E1. (**DCLICKL** on the *Main* icon, if necessary.)

2. **DCLICKL** on the *Schematics* icon to open the *Schematics* window of Figure E2.

3. To change the size of the window, first realize that all windows can be in one of three size modes: full, icon, or intermediate-changeable. The mode is selected by the two size *buttons* at the upper right of the window, which can be in either of the states of Figure E3. Note the state your window is presently in.

4. In either case, **CLICKL** on left-hand downward-pointing triangle and reduce the window to icon mode (see icon at bottom of screen).

5. **DCLICKL** on *Schematics* icon to return to previous size mode.

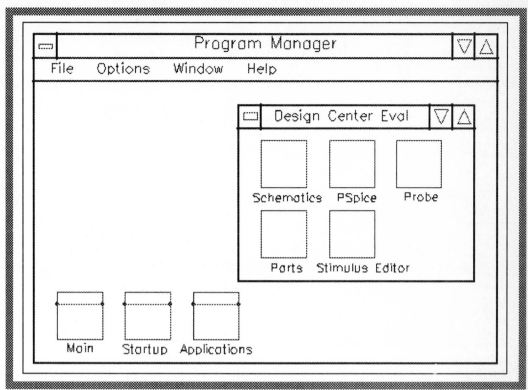

Figure E1   The Program Manager Window

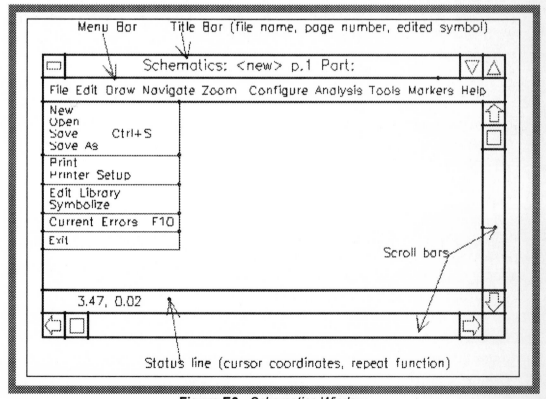

Figure E2   *Schematics Window*

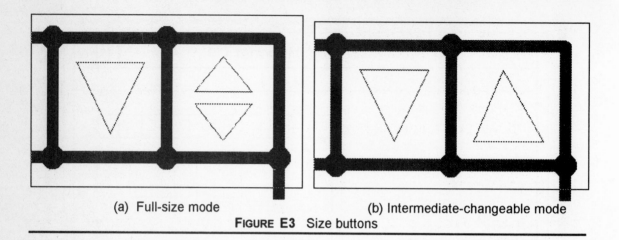

(a) Full-size mode   (b) Intermediate-changeable mode

FIGURE E3   Size buttons

6. If your window's size buttons look like Figure E3(b), you are in full-size mode and the window fills the entire screen.   If the size buttons look like Figure E3(a), you are in intermediate changeable-size mode and the window can be any size.   **CLICKL** on right-hand button several times and note how the window toggles between full and intermediate mode.   When done experimenting with this feature, leave the window in intermediate mode [Figure E3(b)].

7. To change the window to any size, move cursor to left edge of window and note how cursor changes into dual arrows.   **CLICKLH** and drag left-hand window edge back and forth.   When desired size of window has been set, release left mouse button.

8. Repeat step 7 for right-hand, top, and bottom sides of window.

9. Move cursor to upper left corner of window, **CLICKLH,** drag window to desired size, release left-hand mouse button.

10.   Repeat step 9 for other corners of window.

11. Return to Figure E2 and note *Menu Bar* beneath the *Title Bar.*   These are the *Main Menu* selections.

12. To open up the *File* submenu, **File** (**CLICKL** on *File*).   Note the File submenu, as shown in Figure E2.

13. **CLICKL** on "open" to bring up *Open Dialog Box.*   Normally we would enter information into the dialog box, followed by **OK**.   This time, just **Cancel**.

14. **Draw**, **Get New Part** to open up "Draw" dialog box.   Enter "r" from keyboard, **OK**, **CLICKL**, **CLICKR** to place resistor symbol.

15. To move data *within* the window (rather than moving the entire window), **CLICKL** on horizontal and vertical Scroll Bars (see Figure E2).   Note how resistor moves about the screen in large steps.

16. **CLICKL** on arrows at ends of scroll bars and note how the resistor moves in small steps.

17. **CLICKLH** on square button within scroll bar and drag up and down.   Note movement of resistor.

18. Done with this Window's tutorial, **File**, **Exit**.

Note: when more than one window is open, it is easy for windows to be "covered up" by other windows.   To bring each window to the forefront in sequence, press ALT-ESC continuously.

For complete information on the use of Windows, consult the Windows User's Guide from Microsoft Corp.

Index